中国水旱灾害防御公报

China Flood and Drought Disaster Prevention Bulletin

2023

中华人民共和国水利部

Ministry of Water Resources of the People's Republic of China

中国水利水电出版社
www.waterpub.com.cn

·北京·

图书在版编目（CIP）数据

中国水旱灾害防御公报. 2023 / 中华人民共和国水利部编著. -- 北京：中国水利水电出版社，2024.5
ISBN 978-7-5226-2450-1

Ⅰ.①中… Ⅱ.①中… Ⅲ.①水灾－灾害防治－公报－中国－2023②干旱－灾害防治－公报－中国－2023 Ⅳ.①P426.616

中国国家版本馆CIP数据核字(2024)第091246号

责任编辑：徐丽娟

审图号：GS京(2024)0818号

书　　名	中国水旱灾害防御公报2023 ZHONGGUO SHUIHAN ZAIHAI FANGYU GONGBAO 2023
作　　者	中华人民共和国水利部
出版发行	中国水利水电出版社 （北京市海淀区玉渊潭南路1号D座　100038） 网址：www.waterpub.com.cn E-mail：sales@mwr.gov.cn 电话：（010）68545888（营销中心）
经　　售	北京科水图书销售有限公司 电话：（010）68545874、63202643 全国各地新华书店和相关出版物销售网点
排　　版	北京水精灵教育科技有限公司
印　　刷	天津联城印刷有限公司
规　　格	210mm×285mm　16开本　8.125印张　250千字
版　　次	2024年5月第1版　2024年5月第1次印刷
定　　价	89.00元

凡购买我社图书，如有缺页、倒页、脱页的，本社发行部负责调换

版权所有·侵权必究

《中国水旱灾害防御公报》编委会

Editorial Board of *China Flood and Drought Disaster Prevention Bulletin*

主　　任：刘伟平
副 主 任：姚文广　彭　静
成员单位：各省、自治区、直辖市水利（水务）厅（局），新疆生产建设兵团
　　　　　水利局，水利部各流域管理机构，中国水利水电科学研究院

Director: Liu Weiping
Deputy Directors: Yao Wenguang, Peng Jing
Member Units: The Water Resources Departments of the provinces/autonomous regions/municipalities directly under the Central Government, the Water Resources Department of Xinjiang Production and Construction Corps, River Basin Commissions of the Ministry of Water Resources, and China Institute of Water Resources and Hydropower Research

《中国水旱灾害防御公报》编辑部

Editorial Office of *China Flood and Drought Disaster Prevention Bulletin*

主　　编：杨卫忠　吕　娟
副 主 编：吴泽斌　杨　昆　李　岩
成　　员：（以姓氏拼音为序）
　　　　　范　瑱　黄　慧　火传鲁　江　密　李琛亮　李德智　李开峰
　　　　　李铁光　李晓晨　凌永玉　路江鑫　罗　鹏　毛　艳　苗世超
　　　　　苗　雪　穆　磊　谭尧耕　田亚男　王凤恩　王　洁　王　恺
　　　　　许田柱　闫永銮　姚力玮　赵乐媛　朱春子
英文翻译：孟　圆　李若曦

Chief Editors: Yang Weizhong, Lyu Juan
Associate Editors: Wu Zebin, Yang Kun, Li Yan
Editors: (in order of surname pinyin)
Fan Zhen, Huang Hui, Huo Chuanlu, Jiang Mi, Li Chenliang, Li Dezhi, Li Kaifeng
Li Tieguang, Li Xiaochen, Ling Yongyu, Lu Jiangxin, Luo Peng, Mao Yan, Miao Shichao
Miao Xue, Mu Lei, Tan Yaogeng, Tian Ya'nan, Wang Feng'en, Wang Jie, Wang Kai
Xu Tianzhu, Yan Yongluan, Yao Liwei, Zhao Leyuan, Zhu Chunzi
English Translators: Meng Yuan, Li Ruoxi

CONTENTS 目录

1 雨水情 RAINFALL AND WATER REGIME — 1
- 1.1 雨情 / Rainfall — 2
- 1.2 水情 / Water Regime — 4

2 洪涝灾害防御 FLOOD DISASTER PREVENTION — 12
- 2.1 汛情 / Floods — 13
- 2.2 主要江河洪水过程 / Major Flood Processes — 20
- 2.3 典型山洪灾害事件 / Major Flash Flood Disasters — 28
- 2.4 洪涝灾情 / Disasters and Losses — 30
- 2.5 防御工作 / Prevention and Control — 43
- 2.6 防御成效 / Effectiveness of Flood Disaster Prevention — 62

3 干旱灾害防御 DROUGHT DISASTER PREVENTION — 72
- 3.1 旱情 / Droughts — 73
- 3.2 主要干旱过程 / Major Drought Processes — 74
- 3.3 干旱灾情 / Disasters and Losses — 78
- 3.4 防御工作 / Prevention and Control — 86
- 3.5 防御成效 / Effectiveness of Drought Disaster Prevention — 97

4 基础工作 FOUNDATIONAL WORK — 104

- 4.1 机构职能 / Institutional Functions — 105
- 4.2 规章制度 / Rules and Regulations — 105
- 4.3 方案预案 / Contingency Planning — 106
- 4.4 山洪灾害防治 / Flash Flood Disaster Management — 108
- 4.5 蓄滞洪区建设管理 / Construction and Management of Flood Detention Areas — 110
- 4.6 信息发布 / Information Dissemination — 112
- 4.7 复盘分析 / Review and Analysis — 114

附录 APPENDIX

1950—2023 年全国水旱灾情统计
STATISTICS OF FLOOD AND DROUGHT DISASTERS IN CHINA 1950-2023 — 116

　　2023年，我国气候极端反常，部分地区暴洪急涝、旱涝急转、情势偏重。海河流域发生流域性特大洪水，松花江部分支流发生超历史实测记录洪水，秋季洪水多发，西南、西北部分地区发生1961年以来最严重干旱，水旱灾害防御形势异常复杂严峻。党中央、国务院高度重视防汛抗旱工作，习近平总书记多次主持召开会议研究水利工作，亲自部署、亲自指挥防汛抗洪救灾工作，专程考察、专题研究灾后恢复重建工作。李强总理等领导同志多次专题研究或批示解决重大问题。水利部坚决贯彻习近平总书记重要指示精神，按照党中央、国务院决策部署，坚持人民至上、生命至上，树牢底线思维、极限思维，组织各级水利部门、会同有关部门和地方，全力以赴夺取水旱灾害防御全面胜利。

注
（1）本公报香港特别行政区、澳门特别行政区和台湾省统计数据暂缺，新疆生产建设兵团统计数据计入新疆维吾尔自治区统计数据；
（2）本公报所采用的计量单位部分沿用水利统计惯用单位，未进行换算；
（3）本公报数据来源于水利部、应急管理部，水文数据依据水利部信息中心业务系统数据，部分数据未经整编，未注明来源的数据均来源于水利部，指标解释分别参阅《水旱灾害防御统计调查制度（试行）（2021）》和《自然灾害情况统计调查制度（2020）》。

In 2023, the weather pattern was extremely unusual in China. Parts of the country were lashed with rainstorm-induced floods with sudden onset, acute transitions from too much water to too less, and general severity of water disasters. The Haihe River Basin saw basin-wide extreme floods, some tributaries of the Songhua River experienced record-breaking floods, the autumn season saw the most flood events, and parts of Southwest and Northwest China endured the worst drought spell since 1961.

The central government of China was committed to the flood control and drought relief work: President Xi Jinping held consultation meetings on water conservancy, flood control and drought relief and made clear instructions. The President also was attentive to the post-disaster recovery and restoration work. Premier Li Qiang showed equal attention to the details of implementation. The Ministry of Water Resources (hereinafter MWR) thoroughly and forcefully executed the key instructions of President Xi Jinping on flood prevention and drought relief, and the decisions and deployment of the CPC Central Committee and the State Council, and put the safety of people's lives in the first place. Water resources departments at all levels were commanded and guided by the Ministry, in consultation with other relevant departments and local governments, to tighten up the nerves against flood and drought disasters, and plan in details and prepare for the worst. The fight against flood and drought disasters in 2023 was successful thanks to the all-out efforts.

Notes

(1) The data in this Bulletin does not include statistics of the Hong Kong Special Administrative Region (SAR), the Macao Special Administrative Region (SAR) and Taiwan, and the statistics of the Xinjiang Production and Construction Corps is included in the statistics of the Xinjiang Uygur Autonomous Region;

(2) The units of measurement used in this Bulletin conform to what are customarily used in water conservancy and are not converted;

(3) The data in this Bulletin are from the Ministry of Water Resources (MWR) and the Ministry of Emergency Management (MEM), the data on precipitation are drawn upon the flood data reported on the service platform of MWR Water Resources Information Center, the data that do not indicate the source are all from the MWR, and interpretations on the indicators can be found in *Statistical Investigation System for Flood and Drought Disaster Prevention (Trial)(2021)* and *Statistical Investigation System for Natural Disasters (2020)*, respectively.

1 雨水情
RAINFALL AND WATER REGIME

1.1 雨情

2023年，全国平均降水量600毫米，较常年（625毫米）偏少4%，共发生35次强降水过程；与常年相比，海河、淮河、松辽流域降水量偏多1～2成，黄河流域略偏多，长江、珠江及太湖流域（片）略偏少。5—9月，全国平均降水量437毫米，较常年同期（453毫米）偏少4%，其中淮河、海河流域偏多1～3成，松辽、太湖流域（片）略偏多，长江、珠江流域略偏少。全国雨情有3个主要特点。

（1）空间上东北黄淮海偏多。东北中部、黄淮海大部、西北东南部等地降水量较常年偏多2～5成，其中河南省、黑龙江省降水量较常年分别偏多26%、19%，为1961年有完整资料以来第5多、第6多。江南、西南、华南西部、西北中部等地降水量较常年偏少2～5成，其中贵州省、云南省降水量较常年分别偏少22%、18%，为1961年以来第2少、第4少。

（2）时间上前少后多。1—6月，全国平均降水量240毫米，较常年同期（277毫米）偏少13%，其中1月、5月分别偏少36%、18%。7—12月，全国平均降水量360毫米，较常年同期（349毫米）偏多3%，其中8月偏多6%、11月偏多12%。

（3）台风降水强度大。受2305号台风"杜苏芮"影响，海河流域7月28日至8月1日平均降水量155毫米，累计降水量达494亿立方米，为1963年以来最多，最大点降水量北京门头沟区清水站1014.5毫米。受"杜苏芮"残留云系和2306号台风"卡努"影响，松花江支流拉林河流域8月1—5日平均降水量170毫米，占汛期总降水量的43%。受2311号台风"海葵"影响，9月7—8日，广东深圳市、佛山市、肇庆市日降水量突破当地历史实测记录。

1.1 Rainfall

In 2023, the average annual precipitation in China was 600 mm, 4% less than normal (625 mm). A total of 35 heavy rainfall processes occurred. Compared with normal years, the Haihe, Huaihe, and Songhua-Liaohe river basins received 10%-20% more rainfall than normal; the Yellow River basin received slightly more; and the Yangtze, the Pearl, and the Taihu Lake basins got slightly less. From May to September, the national average precipitation was 437 mm, 4% less than normal over the same period (453 mm); the Huaihe River and Haihe River basins received 10%-30% more; the Songhua-Liaohe and the Taihu Lake basins received slightly more; and the Yangtze River and the Pearl River basins received slightly less.

In general, rainfall in 2023 took on the following three characteristics:

(1) More precipitation was in the Northeast and the Huang-Huaihe-Haihe regions. The precipitation in central Northeast China, the vast parts of Huang-Huai-Hai region, the southeast of Northwest China was 20%-50% more than normal; in particular, the annual precipitation in Henan and Heilongjiang provinces was 26% and 19% more than normal, respectively, which was the fifth and sixth highest since the complete documentation was available in 1961. Jiangnan (areas south of the middle-lower Yangtze River), Southwest China, the west of South China, and the central of Northwest China received 20%-50% less than moral; in particular, precipitation in Guizhou and Yunnan provinces was 22% and 18% less than normal, respectively, which is the second and fourth lowest since 1961.

(2) More precipitation was in the latter half of the year than in the first half. From January to June, the national average precipitation was 240 mm, 13% less than normal over the same period (277 mm); the precipitation in January and May was 36% and 18% less than normal, respectively. From July to December, the national average precipitation was 360 mm, 3% more than normal (349 mm); the precipitation in August and November was 6% and 12% more than their respective normals.

(3) Intense precipitation was brought by typhoons. Affected by Typhoon Doksuri (No. 2305), Haihe River basin received an average precipitation of 155 mm from July 28 to August 1, the cumulative precipitation reached a post-1963 new high of 49.4 billion m^3, and the largest point precipitation was recorded at 1,014.5 mm at Qingshui Station in Beijing's Mentougou District. Affected by the residual cloud system of "Doksuri" and Typhoon "Khanun" (No. 2306), the Lalin River basin that drains to the Songhua River received an average precipitation of 170 mm during August 1 to 5, accounting for 43% of the total precipitation in the flood season. Affected by Typhoon "Haikui" (No. 2311) the daily precipitation in Shenzhen, Foshan and Zhaoqing cities of Guangdong Province broke the local historical records from September 7 to 8.

1.2 水情

1.2.1 江河径流量

2023年，全国主要江河径流量较常年总体偏少。长江大通站6553.0亿立方米，偏少3成。黄河花园口站305.4亿立方米，偏少1.5成。淮河鲁台子站174.6亿立方米，偏少1成。海河流域永定河三家店站4.66亿立方米，偏多12.3倍；拒马河张坊站13.9亿立方米，偏多9.7倍；子牙河北中山站14.31亿立方米，偏多14倍。珠江流域西江梧州站1128.0亿立方米，偏少5成；北江石角站346.6亿立方米，偏少2成；东江博罗站162.7亿立方米，偏少3成。松花江佳木斯站764.3亿立方米，偏多5成。辽河铁岭站38.6亿立方米，偏多3成。

汛前，主要江河径流量及与常年同期相比，长江大通站1429.0亿立方米，偏少2成。黄河花园口站92.3亿立方米，偏少1成。淮河鲁台子站11.8亿立方米，偏少6成。海河流域永定河三家店站1.56亿立方米，偏多12倍；拒马河张坊站0.2亿立方米，偏少1成；子牙河北中山站2.57亿立方米，偏多41.8倍。珠江流域西江梧州站237.0亿立方米，偏少2成；北江石角站111.0亿立方米，略偏多；东江博罗站44.0亿立方米，偏少2成。松花江佳木斯站124.0亿立方米，偏多1成。辽河铁岭站8.0亿立方米，偏多7成。

注 松花江、辽河、海河、黄河流域径流量统计时段划分：汛前（1—5月）、汛期（6—9月）、汛后（10—12月）；淮河、长江、珠江流域径流量统计时段划分：汛前（1—4月）、汛期（5—9月）、汛后（10—12月）；太湖暂不做径流量统计。

1.2 Water Regime

1.2.1 River discharge

In 2023, the discharge of major rivers in China was less than normal. Datong Station on the Yangtze River documented 655.3 billion m^3, 30% less than normal; Huayuankou Station on the Yellow River documented 30.54 billion m^3, 15% less; Lutaizi Station on the Huaihe River got 17.46 billion m^3, 10% less; Sanjiadian Station on the Yongding River (draining to Haihe River basin) got 466 million m^3, 12.3 times more; Zhangfang Station on the Juma River got 1.39 billion m^3, 9.7 times more, Beizhongshan Station on Ziya River got 1.431 billion m^3, 14 times more; in the Pearl River Basin, Wuzhou Station on Xijiang River received 112.80 billion m^3, 50% less; Shijiao Station on Beijiang River received 34.66 billion m^3, 20% less; Boluo Station on Dongjiang River received 16.27 billion m^3, 30% less; Jiamusi Station on Songhua River received 76.43 billion m^3, 50% more; Tieling Station on Liaohe River received 3.86 billion m^3, 30% more.

In terms of the discharge of major rivers prior to the flood season, Datong Station on the Yangtze River documented 142.90 billion m^3, 20% less than normal; Huayuankou Station on the Yellow River 9.23 billion m^3, 10% less; Lutaizi Station on Huaihe River 1.18 billion m^3, 60% less. Sanjiadian Station on Yongding River (draining to Haihe River basin) 156 million m^3, 12 times more; Zhangfang Station on the Juma River 20 million m^3, 10% less; Beizhongshan Station on Ziya River 257 million m^3, 41.8 times more; in the Pearl River Basin, Wuzhou Station on Xijiang River 23.7 billion m^3, 20% less; Shijiao Station on Beijiang River 11.10 billion m^3, slightly more; Boluo Station on Dongjiang River 4.40 billion m^3, 20% less; Jiamusi Station on Songhua River 12.40 billion m^3, 10% more; Tieling Station on Liaohe River 800 million m^3, 70% more.

Note For statistics of river discharges in the Songhua River, the Liaohe River, the Haihe River, and the Yellow River basins: pre-flood period (January-May), flood period (June-September), post-flood period (October-December); for statistics of river discharges in the Huaihe, the Yangtze and the Pearl River basins: pre-flood period (January-April), flood period (May-September), post-flood period (October-December); No statistics of discharges in the Taihu Lake are prepared.

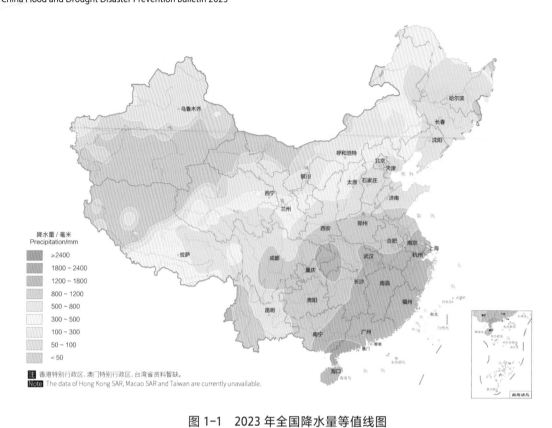

图 1-1　2023 年全国降水量等值线图
Figure 1-1　Isogram of national precipitation in 2023

图 1-2　2023 年全国降水量距平图
Figure 1-2　National precipitation anomalies in 2023

1 雨水情

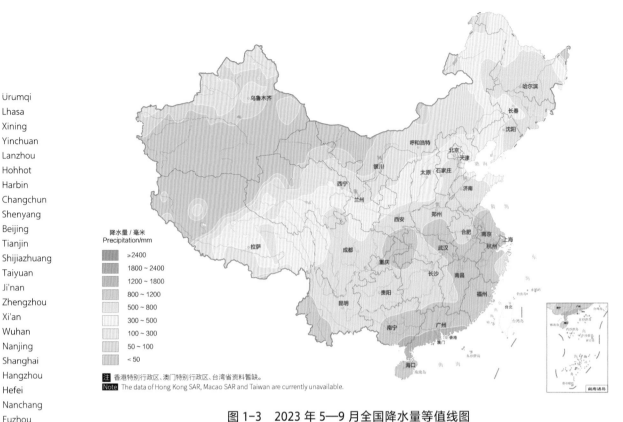

图1-3　2023年5—9月全国降水量等值线图

Figure 1-3　Isogram of precipitation from May to September 2023

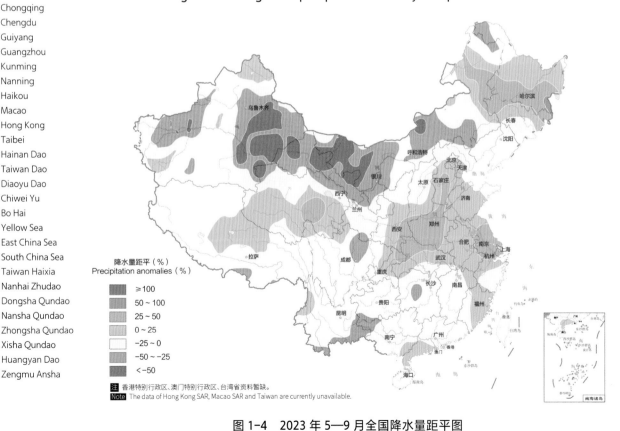

图1-4　2023年5—9月全国降水量距平图

Figure 1-4　Precipitation anomalies from May to September 2023

-7-

汛期，主要江河径流量及与常年同期相比，长江大通站 3515.0 亿立方米，偏少 4 成。黄河花园口站 153.0 亿立方米，偏少 1 成。淮河鲁台子站 108.0 亿立方米，偏少 2 成。海河流域永定河三家店站 2.64 亿立方米，偏多 21 倍；拒马河张坊站 12.6 亿立方米，偏多 14 倍；子牙河北中山站 9.49 亿立方米，偏多 12 倍。珠江流域西江梧州站 703.0 亿立方米，偏少 5 成；北江石角站 197.0 亿立方米，偏少 3 成；东江博罗站 86.4 亿立方米，偏少 4 成。松花江佳木斯站 518.3 亿立方米，偏多 7 成。辽河铁岭站 26.4 亿立方米，偏多 2 成。

汛后，主要江河径流量及与常年同期相比，长江大通站 1609.0 亿立方米，偏少 1 成。黄河花园口站 60.1 亿立方米，偏少 3 成。淮河鲁台子站 54.8 亿立方米，偏多 8 成。海河流域永定河三家店站 0.46 亿立方米，偏多 2.8 倍；拒马河张坊站 1.1 亿立方米，偏多 3 倍；子牙河北中山站 2.24 亿立方米，偏多 13 倍。珠江流域西江梧州站 188.0 亿立方米，偏少 3 成；北江石角站 38.6 亿立方米，偏少 2 成；东江博罗站 32.3 亿立方米，偏少 1 成。松花江佳木斯站 122.0 亿立方米，偏多 2 成。辽河铁岭站 4.2 亿立方米，偏多 3 成。

In terms of the discharge of major rivers during the flood season, Datong Station on the Yangtze documented 351.50 billion m^3, 40% less than normal; Huayuankou Station on the Yellow River 15.30 billion m^3, 10% less; Lutaizi Station on Huaihe 10.8 billion m^3, 20% less; Sanjiadian Station on the Yongding River (draining to Haihe) 264 million m^3, 21 times more; Zhangfang Station on the Juma River 1.26 billion m^3, 14 times more; Beizhongshan Station on Ziya 949 million m^3, 12 times more; in the Pearl River Basin, Wuzhou Station on Xijiang 70.30 billion m^3, 50% less; Shijiao Station on Beijiang 19.70 billion m^3, 30% less; Boluo Station on Dongjiang 8.64 billion m^3, 40% less; Jiamusi Station on Songhua 51.83 billion m^3, 70% more; Tieling Station on Liaohe 2.64 billion m^3, 20% more.

In terms of the discharge of major rivers after the flood season, Datong Station on the Yangtze documented 160.90 billion m^3, 10% less than normal; Huayuankou Station on the Yellow River 6.01 billion m^3, 30% less; Lutaizi Station on Huaihe 5.48 billion m^3, 80% more; Sanjiadian Station on the Yongding River (draining to Haihe) 46 million m^3, 2.8 times more; Zhangfang Station on the Juma River 110 million m^3, 3 times more; Beizhongshan Station on Ziya 224 million m^3, 13 times more; in the Pearl River Basin, Wuzhou Station on Xijiang 18.80 billion m^3, 30% less; Shijiao Station on Beijiang 3.86 billion m^3, 20% less; Boluo Station on Dongjiang 3.23 billion m^3, 10% less; Jiamusi Station on Songhua 12.20 billion m^3, 20% more; Tieling Station on Liaohe 420 million m^3, 30% more.

1.2.2 水库蓄水

6月1日，纳入水利部日常统计范围的 8238 座水库（以下简称"统计水库"）蓄水量 4133.4 亿立方米，较常年同期偏多 9%。其中，729 座大型水库蓄水量 3727.1 亿立方米，较常年同期偏多 10%；3144 座中型水库蓄水量 367.1 亿立方米，较常年同期偏少 3%；4365 座小型水库蓄水量 39.2 亿立方米，较常年同期偏多 7%。

10月1日，统计水库蓄水量 5362.5 亿立方米，较 6 月 1 日增加 30%，较常年同期偏多 6%。其中，729 座大型水库蓄水量 4863.3 亿立方米，较 6 月 1 日增加 30%，较常年同期偏多 6%；3144 座中型水库蓄水量 454.1 亿立方米，较 6 月 1 日增加 24%，较常年同期偏多 5%；4365 座小型水库蓄水量 45.1 亿立方米，较 6 月 1 日增加 15%，较常年同期偏多 17%。

年末，统计水库蓄水量 5184.1 亿立方米，较 1 月 1 日增加 8%，较常年同期偏多 9%。其中，729 座大型水库蓄水量 4711.8 亿立方米，较 1 月 1 日增加 7%，较常年同期偏多 10%；3144 座中型水库蓄水量 431.7 亿立方米，较 1 月 1 日增加 13%，较常年同期偏多 3%；4365 座小型水库蓄水量 40.6 亿立方米，较 1 月 1 日增加 11%，较常年同期偏少 24%。

表 1-1 2023 年统计水库蓄水量情况
Table 1-1 Water storage of the daily-reporting reservoirs in 2023

时间 Date	8238 座水库蓄水量 / 亿立方米 Storage in 8,238 reservoirs / 100 million m³			
	729 座大型水库 729 large reservoirs	3144 座中型水库 3,144 medium-sized reservoirs	4365 座小型水库 4,365 small reservoirs	合计 Total
1月1日 January 1	4391.2	381.0	36.7	4808.9
6月1日 June 1	3727.1	367.1	39.2	4133.4
10月1日 October 1	4863.3	454.1	45.1	5362.5
年末 Year-end	4711.8	431.7	40.6	5184.1

注 年末数据指 2024 年 1 月 1 日 8 时统计数据。
Note Year-end data as of 8:00 am on January 1, 2024.

1.2.2 Reservoir storage

On June 1, the 8,238 reservoirs that report daily statistics to MWR (hereinafter referred to as daily-reporting reservoirs) had a total storage of 413.34 billion m^3, 9% more than normal over the same period. Among them, the 729 large reservoirs had a water storage of 372.71 billion m^3, 10% more than normal; the 3,144 medium-sized reservoirs had a total storage of 36.71 billion m^3, 3% less than normal; and the 4,365 small reservoirs had a total water storage of 3.92 billion m^3, 7% more than normal.

On October 1, the storage of daily-reporting reservoirs was 536.25 billion m^3, 30% more than that on June 1 and 6% more than normal over the same period. Among them, the 729 large reservoirs stored 486.33 billion m^3, 30% more than that on June 1 and 6% more than normal; the 3,144 medium-sized reservoirs stored 45.41 billion m^3, 24% more than that on June 1 and 17% more than normal; and the 4,365 small reservoirs stored 4.51 billion m^3, 15% more than that on June 1 and 17% more than normal.

At the end of the year, the storage of daily-reporting reservoirs was 518.41 billion m^3, 8% more than that on January 1 and 9% more than normal over the same period. Among them, the 729 large reservoirs stored 471.18 billion m^3, 7% more than that on January 1 and 10% more than normal; the 3,144 medium-sized reservoirs stored 43.17 billion m^3, 13% more than that on January 1 and 3% more than normal; and the 4,365 small reservoirs stored 4.06 billion m^3, 11% more than that on January 1 and 24% less than normal.

2 洪涝灾害防御
FLOOD DISASTER PREVENTION

2.1 汛情

2023年3月24日，我国进入汛期，较多年平均入汛日期（4月1日）偏早8天。2023年，全国主要江河共发生4次编号洪水，共有708条河流发生超警戒洪水，较常年平均（675条）偏多5%，其中，129条河流发生超保证洪水，49条河流发生超历史实测记录洪水。长江流域（片）有213条河流发生超警戒洪水，其中，重庆嘉陵江水系雍溪河、三峡区间朱衣河等62条河流发生超保证洪水，金沙江沱沱河至岗拖江段、鄱阳湖水系抚河支流曹水等14条河流发生超历史实测记录洪水；黄河流域（片）有36条河流发生超警戒洪水，其中，新疆别列孜克河、古库尔苏河等4条河流发生超保证洪水，新疆克兰河、内蒙古黑河等2条河流发生超历史实测记录洪水；淮河流域（片）有21条河流发生超警戒洪水，其中，江苏大运河、房亭河等2条河流发生超保证、超历史实测记录洪水；海河流域（片）有23条河流发生超警戒洪水，其中，拒马河及其支流白沟河等12条河流发生超保证洪水，永定河及支流清水河等8条河流发生超历史实测记录洪水，子牙河、永定河、大清河分别发生1次编号洪水，永定河发生1924年以来最大洪水，大清河发生1963年以来最大洪水，海河流域发生流域性特大洪水；珠江流域（片）有180条河流发生超警戒洪水，其中，福建汀江支流小澜溪、贵州红水河支流京舟河等9条河流发生超保证洪水，广东西江支流罗定江、义昌江等5条河流发生超历史实测记录洪水；松辽流域（片）有140条河流发生超警戒洪水，其中，松花江支流牡丹江、嫩江支流洮儿河等27条河流发生超保证洪水，松花江支流拉林河、蚂蚁河等18条河流发生超历史实测记录洪水，松花江发生1次编号洪水；太湖流域（片）有95条河流发生超警戒洪水，其中，福建闽南沿海九十九溪、浙江钱塘江上游兰江等13条河流发生超保证洪水。黄河、黑龙江等北方河流凌情平稳。

注 河流超警戒（超保证、超历史）包括水位超警戒（超保证、超历史）或流量超警戒（超保证、超历史）。

2.1 Floods

The 2023 flood season in China began on March 24, eight days earlier than the multi-year average start date (April 1). There were four numbered floods in major rivers across the country in 2023. A total of 708 rivers swelled above the warning level, a 5% increase over the average annual (675 rivers). Among them, 129 experienced floods beyond the guaranteed level and 49 rivers experienced record-breaking floods. In the Yangtze River basin/region, 213 rivers experienced floods beyond the warning level; among them, 62 rivers, including the Yongxi River in the Jialing River system in Chongqing and the Zhuyi River in the Three Gorges region, had floods beyond the guaranteed level; and 14 rivers, including the section from Tuotuo River to Gangtuo River of the Jinsha River and the Caoshui River, a tributary of the Fuhe River that drains to the Poyang Lake system, had record-breaking floods. In the Yellow River basin/region, 36 rivers experienced floods beyond the warning levels; among them, 4 rivers such as the Bieliezike River and the Guku'ersu River in Xinjiang had floods beyond the guaranteed level, and two rivers, namely the Kelan River in Xinjiang and Heihe River in Nei Mongol, had record-breaking floods. In the Huaihe River basin/region, 21 rivers experienced floods beyond the warning level; among them 2 rivers, namely the Dayun River and the Fangting River in Jiangsu, had floods beyond the guaranteed level and record-breaking floods, respectively. In the Haihe River basin/region, 23 rivers experienced floods beyond the warning levels; among them, 12 rivers such as the Juma River and its tributary

---

Note   River floods that break the warning/guaranteed/maximum measured levels include those that break the warning/guaranteed/maximum measured water level and those that break the warning/guaranteed/maximum measured discharges.

Baigou had floods beyond the guaranteed level, and 8 rivers, including the Yongding River and its tributary Qingshui, had record-breaking floods. The Ziya, Yongding and Daqing rivers each withstood one numbered flood. The Yongding River experienced its largest flood since 1924, the Daqing River experienced its largest flood since 1963, and the Haihe River Basin experienced an extreme basin-wide flood.In the Pearl River basin/region, 180 rivers experienced floods beyond the warning level; among them 9 rivers, including the Xiaolanxi River, a tributary of the Tingjiang River in Fujian, and the Jingzhou River, a tributary of the Hongshui River in Guizhou, had floods beyond the guaranteed level; and 5 rivers such as the Luoding and Yichang rivers, tributaries of Xijiang in Guangdong, had record-breaking floods. In the Songhua-Liaohe River basin/region, 140 rivers experienced floods beyond the warning level; among them, 27 rivers such as the Mudan River, a tributary of the Songhua River, and the Tao'er River, a tributary of Nenjiang, had floods beyond the guaranteed level; and 18 rivers, including the Lalin and Mayi rivers, tributaries of the Songhua River, had record-breaking floods. One numbered flood occurred in the Songhua River. In the Taihu Lake basin/region, 95 rivers experienced floods beyond the warning level; among them, 13 rivers such as the Jiushijiuxi in the southern coastal Fujian and the Lanjiang River draining to the upper Qiantang River in Zhejiang, had floods beyond the guaranteed level. China's northern rivers, including the Yellow River and the Heilong River, safely passed the ice flood period.

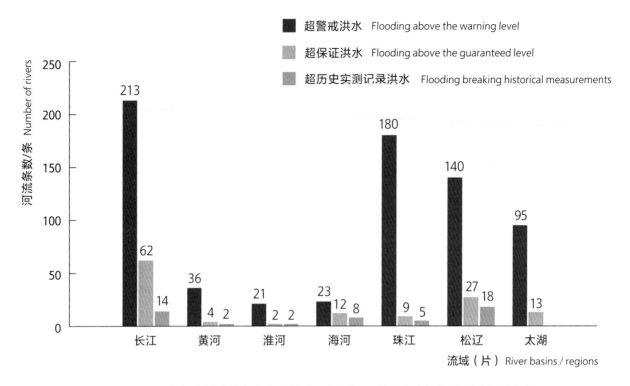

图 2-1　2023 年各流域（片）发生超警戒、超保证、超历史实测记录洪水河流条数

Figure 2-1　Number of rivers experiencing floods above the warning water level, above the guaranteed water level, and breaking historical measurements in the major river basins / regions in 2023

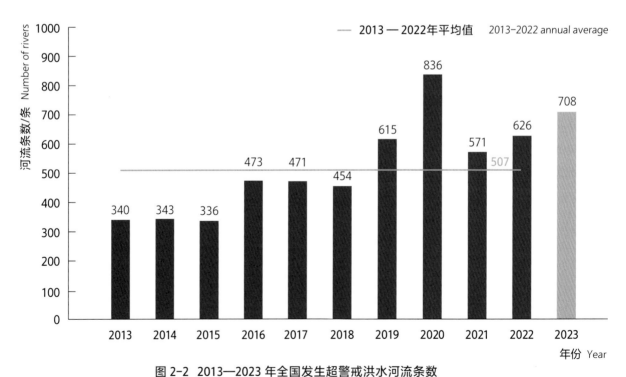

图 2-2　2013—2023 年全国发生超警戒洪水河流条数

Figure 2-2　Number of rivers experiencing floods above the warning level in China 2013-2023

2 洪涝灾害防御

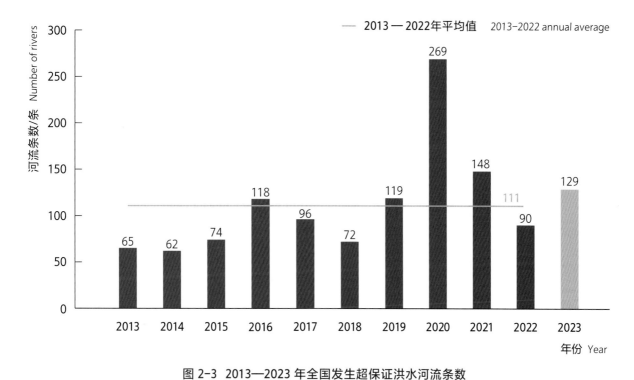

图 2-3 2013—2023 年全国发生超保证洪水河流条数
Figure 2-3 Number of rivers experiencing floods above the guaranteed level in China 2013-2023

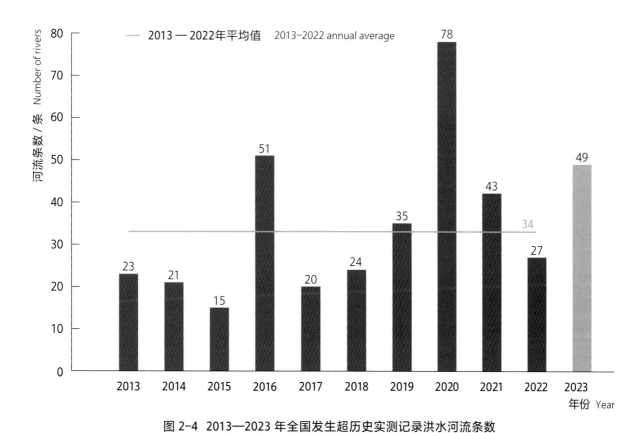

图 2-4 2013—2023 年全国发生超历史实测记录洪水河流条数
Figure 2-4 Number of rivers experiencing floods breaking historical measurements in China 2013-2023

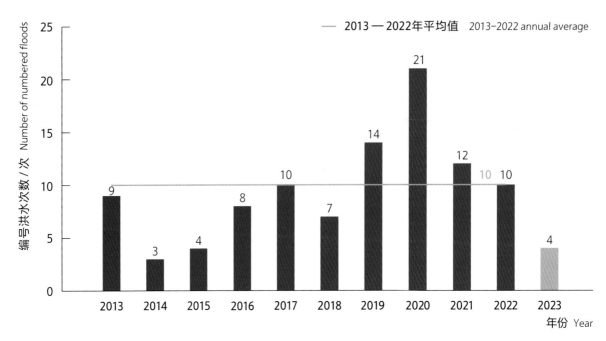

图 2-5　2013—2023 年全国主要江河发生编号洪水次数
Figure 2-5　Statistics of numbered floods in major rivers in China 2013-2023

2023 年全国汛情有 4 个主要特点。

（1）海河洪水涨势猛、量级大。7 月 30 日 23 时至 31 日 11 时（12 小时），子牙河、永定河、大清河相继发生编号洪水。拒马河都衙站流量 1 小时内从 600 立方米每秒猛涨至洪峰流量 5500 立方米每秒，永定河卢沟桥枢纽流量 1.5 小时内从 1000 立方米每秒猛涨至 4650 立方米每秒。海河流域发生 60 年来最大流域性特大洪水，永定河发生 1924 年以来最大洪水，大清河发生 1963 年以来最大洪水。

（2）松花江部分支流汛情重。7 月底至 8 月底，松花江支流拉林河、蚂蚁河及嫩江支流雅鲁河、洮儿河等 96 条河流超警戒，其中松花江干流发生编号洪水，支流牡丹江超保证，支流拉林河、蚂蚁河全线超历史实测记录，乌苏里江干流虎头站超警戒 37 天，超保证 12 天。

（3）秋季洪水多发。9—10 月，湖北、福建、广东、广西等地有 226 条河流发生超警戒以上洪水，占全年超警戒河流总条数的 32%；汉江发生 2 次编号洪水，中下游干流宜城以下江段水位全线超警戒；西藏发生降水融雪融冰混合型洪水，色林错等高原湖泊出现局部漫溢。

（4）山洪灾害多发重发。受短时局地强降水影响，2023 年山洪灾害呈现多发频发重发态势，北京、河北、四川、重庆、甘肃、黑龙江等地发生多起山洪（或伴生泥石流）灾害，导致多人死亡或失踪。

注　永定河卢沟桥枢纽位于永定河卢沟桥站上游，卢沟桥枢纽过闸洪峰流量 4650 立方米每秒为推算所得，卢沟桥站洪峰流量 4540 立方米每秒为实测值。

Floods that occurred in China in 2023 generally took on the following four characteristics:

(1) The Haihe River was hit by violently surging and voluminous floods. Within the 12 hours from 23:00 on July 30 to 11:00 on July 31, the Ziya, Yongding and Daqing rivers were hit by numbered floods successively. The flow at the Duya Station on the Juma River soared from 600 m^3/s to the peak of 5,500 m^3/s within an hour, and the flow at the Lugouqiao engineering complex on the Yongding River soared from 1,000 m^3/s to 4,650 m^3/s within 1.5 hours. The Haihe River basin experienced its largest basin-wide extreme flood in 60 years, the Yongding River experienced its largest flood since 1924, and the Daqing River experienced its largest flood since 1963.

(2) Several tributaries of the Songhua River withstood severe floods. From the end of July to the end of August, a total of 96 rivers, including the Lalin and Mayi rivers that drain to the Songhua River, and the Yalu and Tao'er rivers that drain to the Nenjiang River, experienced floods beyond the warning level. Among them, the mainstream Songhua River experienced numbered floods, the tributary Mudan experienced floods beyond the guaranteed level, and the tributaries Lalin and Mayi had record-breaking floods over the entire routes. The Hutou Station on the mainstream Wusuli River experienced floods beyond the warning level for 37 days and beyond the guaranteed level for 12 days.

(3) The autumn of 2023 saw frequent floods. From September to October, 226 rivers in Hubei, Fujian, Guangdong, and Guangxi experienced floods beyond the warning level, accounting for 32% of the national total of rivers exceeding the warning levels throughout the year. The Hanjiang River experienced two numbered floods, with water levels along the entire middle to lower section downstream of Yicheng exceeding warning levels. Xizang experienced floods that were triggered by precipitation, snow melt, and ice melt, with localized flooding in its plateau lakes such as the Selin Co.

(4) Flash floods were frequent and severe. Flash flood disasters were frequent and severe under the influence of localized intense precipitation over short duration. Regions including Beijing, Hebei, Sichuan, Chongqing, Gansu, and Heilongjiang were hit by multiple flash flood disasters (or accompanying mudslides), causing many deaths or missing cases.

Note The Lugouqiao engineering complex is located upstream of the Lugouqiao Station on the Yongding River. The peak flow of 4,650 m^3/s at this project is a calculated value, while the peak flow of 4,540 m^3/s at Lugouqiao Station is a measured value.

2.2 主要江河洪水过程

2.2.1 海河流域性特大洪水

受台风"杜苏芮"残余环流北上、地形抬升和副热带高压的共同影响，7月28日至8月1日，海河流域普降大到暴雨、局部降特大暴雨，流域平均降水量155毫米，其中大清河水系305毫米、子牙河水系186毫米、漳卫河水系182毫米、北三河水系139毫米、永定河水系93毫米、徒骇马颊河水系82毫米、滦河水系38毫米，最大点降水量为永定河水系清水河清水站（北京门头沟区）1014.5毫米，流域累计降水总量达494亿立方米。受强降水影响，海河流域有22条河流发生超警戒洪水，7条河流发生超保证洪水，8条河流发生有实测记录以来最大洪水，大清河、永定河发生特大洪水，子牙河发生大洪水，海河流域发生60年来最大流域性特大洪水。

子牙河水系降水过程集中在7月28—29日，流域平均降水量186毫米，暴雨中心位于岗南水库至黄壁庄水库区间和滏阳河上游，最大点降水量为泜河临城站（河北邢台市）1003毫米。7月30日23时滹沱河黄壁庄水库入库流量涨至3615立方米每秒，形成子牙河2023年第1号洪水。7月31日2时朱庄水库入库洪峰流量7900立方米每秒，列1975年有实测记录以来第2位。7月31日7时黄壁庄水库入库洪峰流量6250立方米每秒，列1959年有实测记录以来第5位。

滹沱河黄壁庄水库泄洪（7月31日）
Huangbizhuang Reservoir on the Hutuo River was discharging floods (July 31)

2.2 Major Flood Processes

2.2.1 The extreme basin-wide flood in the Haihe River basin

Due to the combined effects of the northward residual circulation of Typhoon Doksuri, orographic lift, and the subtropical high pressure, from July 28 to August 1, the Haihe River basin experienced widespread heavy to torrential rains and localized extreme rainstorms. The average precipitation in the basin was 155 mm, with 305 mm in the Daqing River system, 186 mm in the Ziya River system, 182 mm in the Zhangwei River system, 139 mm in the Beisan River system, 93 mm in the Yongding River system, 82 mm in the Tuhaimajia River system, and 38 mm in the Luanhe River system. The maximum point precipitation of 1,014.5 mm was recorded at the Qingshui Station (Mentougou District, Beijing) on the Qingshui River that drains to the Yongding River system; the cumulative precipitation in the basin reached 49.4 billion m^3. Affected by intense precipitation, 22 rivers in the Haihe River basin experienced floods beyond the warning levels, 7 rivers had floods beyond the guaranteed levels, and 8 rivers had record-breaking floods. The Daqing and Yongding rivers were hit by extreme flooding, the Ziya River was hit by a major flooding, and the entire Haihe River Basin witnessed its largest basin-wide extreme flood over the past 60 years.

Precipitation across the Ziya River system mainly occurred on July 28 and 29, with an average basin-wide precipitation of 186 mm. The center of the rainstorm was located between the Gangnan Reservoir and Huangbizhuang Reservoir and the upper stream of the Fuyang River. The maximum point precipitation of 1,003 mm was recorded at the Lincheng Station on the Zhihe River (Xingtai City, Hebei). The No. 1 flood of the Ziya River in 2023 formed at 23:00 on July 30, when the inflow to the Huangbizhuang Reservoir on the Hutuo River rose to 3,615 m^3/s. At 02:00 on July 31, the peak inflow to the Zhuzhuang Reservoir reached 7,900 m^3/s, the second highest since measurements began in 1975. At 07:00 on July 31, the peak inflow to the Huangbizhuang Reservoir reached 6,250 m^3/s, the fifth highest since records began in 1959.

永定河水系降水过程集中在 7 月 30—31 日，流域平均降水量 93 毫米，暴雨中心位于官厅山峡区间，最大点降水量为清水河清水站（北京门头沟区）1014.5 毫米。7 月 31 日 11 时，永定河三家店站流量涨至 622 立方米每秒，形成永定河 2023 年第 1 号洪水。7 月 31 日 14 时，卢沟桥站洪峰流量 4540 立方米每秒，列 1924 年有实测记录以来第 1 位。

大清河水系降水过程集中在 7 月 30—31 日，流域平均降水量 305 毫米，暴雨中心位于大清河北支，最大点降水量为大石河霞云岭站（北京房山区）823 毫米。7 月 31 日 11 时拒马河张坊站流量涨至 1610 立方米每秒，形成大清河 2023 年第 1 号洪水。7 月 31 日 11 时大石河漫水河站洪峰流量 5300 立方米每秒，列 1951 年有实测记录以来第 1 位。7 月 31 日 22 时拒马河张坊站洪峰流量 7330 立方米每秒，列 1952 年有实测记录以来第 2 位。

永定河北京段河道水位快速上涨（8 月 1 日）
The Yongding River swelled violently along its Beijing section (August 1)

Precipitation across the Yongding River system mainly occurred on July 30 and 31, with an average basin-wide precipitation of 93 mm. The center of the rainstorm was located in the Guanting mountain gorge area, with the maximum point precipitation of 1,014.5 mm recorded at the Qingshui Station on the Qingshui River (Mentougou District, Beijing). At 11:00 on July 31, the flow at the Sanjiadian Station on the Yongding River increased to 622 m³/s, forming the No.1 flood of the Yongding River in 2023. At 14:00 on July 31, the peak flow at the Lugouqiao Station rated at 4,540 m³/s, the highest since measurements began in 1924.

Precipitation across the Daqing River system mainly occurred on July 30 and 31, with an average basin-wide precipitation of 305 mm. The center of the rainstorm was located in the northern branch of the Daqing River, with the maximum point precipitation of 823 mm recorded at the Xiayunling Station on the Dashi River (Fangshan District, Beijing). At 11:00 on July 31, the flow at the Zhangfang Station on the Juma River rose to 1,610 m³/s, forming the No. 1 flood of the Daqing River in 2023. At 11:00 on July 31, the peak flow at the Manshuihe Station on the Dashi River reached 5,300 m³/s, the highest since measurements began in 1951. At 22:00 on July 31, the peak flow at the Zhangfang Station on the Juma River reached 7,330 m³/s, the second highest since measurements began in 1952.

大清河洪水经独流减河防潮闸下泄入海（8月9日）
Floodwaters in the Daqing River were discharged into sea through the tidal gate on the Duliujian River (August 9)

2.2.2 松花江编号洪水

受台风"杜苏芮"和"卡努"影响，松花江流域出现 2 次大范围强降水过程，主要集中于嫩江、松花江、乌苏里江、绥芬河等流域，降水落区高度重叠。8 月 1—5 日，拉林河、蚂蚁河、牡丹江流域平均降水量分别为 170 毫米、121 毫米、111 毫米，占汛期总降水量的 43%、20%、30%；8 月 2—11 日，绥芬河流域平均降水量 203 毫米，占汛期总降水量的 55.1%。受强降水影响，松花江下游干流佳木斯站 8 月 7 日 20 时水位涨至警戒水位（79.30 米），形成松花江 2023 年第 1 号洪水；拉林河发生特大洪水，干流各站洪水重现期达到或超过 100 年，上游磨盘山水库、牤牛河及龙凤山水库洪水重现期均达 400 年；蚂蚁河上中游发生特大洪水，尚志段、延寿段洪水重现期分别超 100 年、50 年；牡丹江发生超保证洪水。此外，乌苏里江、绥芬河及嫩江支流雅鲁河发生大洪水。

拉林河沿线农田受淹（8 月 9 日）
Farmlands were inundated along the swelling Lalin River (August 9)

2.2.2 Numbered floods in the Songhua River

Affected by typhoons Doksuri and Khanun, the Songhua River basin experienced two extensive heavy precipitation processes. Precipitation mainly concentrated in the basins of the Nenjiang, Songhua, Wusuli, and Suifen rivers, with a high overlap of precipitation areas. From August 1 to 5, the average precipitation in the Lalin, Mayi, and Mudan river basins were 170 mm, 121 mm and 111 mm, respectively, accounting for 43%, 20%, and 30% of the flood-season total precipitation; from August 2 to 11, the average precipitation in the Suifen River basin was 203 mm, accounting for 55.1% of the flood-season total precipitation. Due to the heavy precipitation, the water level at the Jiamusi Station in the lower Songhua rose to the warning level (79.30 m) at 20:00 on August 7, forming the No.1 flood of the Songhua River in 2023; the Lalin River experienced an extreme flood, with the floods at stations along the mainstream reaching or exceeding 100 years in return periods, and the floods at the Mopanshan Reservoir, Mangniu River, and Longfengshan Reservoir upstream all reaching 400 years in return periods; the upper and middle reaches of the Mayi River experienced extreme floods, with the floods in the Shangzhi and Yanshou sections exceeding 100 years and 50 years in return periods; the Mudan River experienced a flood beyond the guaranteed level. Additionally, the Wusuli River, the Suifen River, and the Nenjiang River's tributary Yalu River experienced major floods.

2.2.3 汉江秋季洪水

8月22日至10月6日，汉江流域共发生7次降水过程，累计降水量348.7毫米，较常年同期偏多120%。其中，上游偏多90%，中下游偏多170%。9月下旬至10月上旬，汉江发生2次编号洪水，9月29日20时汉江上游干流丹江口水库入库流量涨至15100立方米每秒，形成汉江2023年第1号洪水，9月30日4时出现最大入库流量16400立方米每秒，重现期超过5年（秋季洪水序列）；10月2日22时，汉江中游干流皇庄站水位涨至48.02米，超过警戒水位0.02米，形成汉江2023年第2号洪水，10月4日6时出现洪峰流量13900立方米每秒，丹江口至皇庄区间最大7天洪量重现期接近10年（秋季洪水序列）。

秋汛期丹江口水库泄洪（9月30日）
Flood discharge at the Danjiangkou Reservoir in the autumn flood season (September 30)

2.2.3 Autumn floods in the Hanjiang River

From August 22 to October 6, the Hanjiang River basin experienced 7 rainfall processes that created 348.7 mm of cumulative precipitation, 120% more than normal over the same period; among these, precipitation in the upper reaches was 90% above average, while the middle and lower reaches exceeded the average by 170%. From late September to early October, the Hanjiang River experienced two numbered floods. At 20:00 on September 29, the inflow to the Danjiangkou Reservoir on the upper Hanjiang River rose to 15,100 m³/s, forming the No.1 flood of the Hanjiang River in 2023. The maximum inflow reached 16,400 m³/s at 04:00 on September 30, amounting to 5 years in return periods (autumn flood sequence). At 22:00 on October 2, the water level at the Huangzhuang Station in the middle Hanjiang rose to 48.02 m, exceeding the warning level by 0.02 m and forming the No.2 flood of the Hanjiang River in 2023. The peak flow of 13,900 m³/s occurred at 06:00 on October 4, with the maximum 7-day flood volume between Danjiangkou and Huangzhuang nearing 10 years in return periods (autumn flood sequence).

2.3 典型山洪灾害事件

8月20日23时至21日1时，四川凉山金阳县芦稿林河（流域面积201.29平方千米）干流上中游发生短时强降水，流域上游最大1小时点降水量54毫米，最大6小时点降水量118毫米，最大24小时点降水量132毫米，重现期均超100年。芦稿林河流域6小时平均降水量58毫米，重现期接近20年。强降水引发多处崩塌、滑坡、泥石流。8月21日0时10分，山洪泥石流冲毁沿江高速金宁一标段项目部钢筋加工场，加之钢筋加工场汛期违规住人、项目部管理人员不执行转移避险指令、未及时组织涉险人员转移，造成6人死亡、46人失踪、21人受伤。经现场洪痕调查等多种方法综合分析，芦稿林河汇入金沙江河口处的控制断面（发生重大人员伤亡的金宁一标段项目部钢筋加工场河段）洪峰流量270立方米每秒（重现期接近20年）。

沿江高速金宁一标段项目部钢筋加工场河段（8月21日）
The destructed steel bar processing plant at Jining section I of the along river expressway project (August 21)

2.3 Major Flash Flood Disasters

From 23:00 on August 20 to 01:00 on August 21, an intense rainfall within a short period occurred in the upper and middle reaches of the Lugaolin River (drainage area 201.29 km²) in Jinyang County, Liangshan, Sichuan. The maximum one-hour point precipitation in the upper reaches reached 54 mm, the maximum six-hour point precipitation reached 118 mm, and the maximum 24-hour point precipitation reached 132 mm, with return periods all exceeding 100 years. The average rainfall over 6 hours in the Lugaolin River basin reached 58 mm, with the return period nearing 20 years. The heavy rainfall triggered landslides, mudslides and debris flows in multiple sites. At 00:10 on August 21, a flash flood and debris flow destroyed a steel bar processing plant for the Jinning section I of the along river expressway project. The violation of regulations by accommodating people in the plant during flood season, the project management's failure to execute evacuation orders, and the delay in organizing the evacuation of people at risk led to 6 deaths, 46 missing, and 21 injuries. Forensic analysis using flood trace investigation and other methods indicates that the peak flow at the control section where the Lugaolin River flows into the Jinsha River (location of the Jinning section I project that suffered heavy casualties) rated at 270 m³/s (with a return period nearing 20 years).

2.4 洪涝灾情

2.4.1 总体灾情

2023 年，洪涝灾害导致全国 5278.93 万人次受灾，比前 10 年平均值下降 24.7%；309 人死亡失踪，比前 10 年平均值下降 41.6%；13.00 万间房屋倒塌，比前 10 年平均值下降 34.1%；农作物受灾 4633.29 千公顷，比前 10 年平均值下降 30.8%，其中绝收 559.01 千公顷；直接经济损失 2445.75 亿元，比前 10 年平均值上升 10.5%；直接经济损失占当年 GDP 的 0.19%，比前 10 年平均值下降 29.6%。北京、河北、黑龙江、河南 4 省（直辖市）灾情较重，因洪涝直接经济损失 1864.72 亿元，占全国 76.2%。

表 2-1 2023 年全国因洪涝受灾人口、死亡失踪人口及直接经济损失情况
Table 2-1 Population affected, deaths and missing persons, direct economic losses by floods in 2023

地区 Province	受灾人口 / 万人次 Affected population/ 10,000 person-times	死亡失踪人口 / 人 Deaths and missing persons/ person	直接经济损失 / 亿元 Direct economic losses / 100 million RMB
全国 Nationwide	5278.93	309	2445.75
北京 Beijing	131.32	61	637.39
天津 Tianjin	13.24		52.23
河北 Hebei	407.02	50	968.33
山西 Shanxi	60.87	3	41.33
内蒙古 Nei Mongol	68.38	4	39.46
辽宁 Liaoning	14.81		8.55
吉林 Jilin	65.44	22	74.61
黑龙江 Heilongjiang	73.53	25	145.56
上海 Shanghai			
江苏 Jiangsu	3.65		0.80
浙江 Zhejiang	3.70	8	5.52
安徽 Anhui	31.49	1	3.60
福建 Fujian	26.09	8	28.76
江西 Jiangxi	161.17	2	18.46
山东 Shandong	11.02		0.82

续表 Continued

地区 Province	受灾人口 / 万人次 Affected population/ 10,000 person-times	死亡失踪人口 / 人 Deaths and missing persons/ person	直接经济损失 / 亿元 Direct economic losses / 100 million RMB
河南 Henan	2414.57	5	113.44
湖北 Hubei	320.44	1	32.91
湖南 Hunan	166.64		33.28
广东 Guangdong	8.00		3.69
广西 Guangxi	106.93	1	32.51
海南 Hainan	0.79		0.29
重庆 Chongqing	140.14	5	43.28
四川 Sichuan	503.07	52	67.91
贵州 Guizhou	94.08	2	10.24
云南 Yunnan	158.64	6	24.40
西藏 Xizang	10.69	2	2.03
陕西 Shanxi	232.07	32	42.26
甘肃 Gansu	46.21	18	10.85
青海 Qinghai	4.61		2.42
宁夏 Ningxia			
新疆 Xinjiang	0.32	1	0.82

注 数据来源于应急管理部，空白表示无灾情，部分地区直接经济损失未完全包括水利工程设施直接经济损失。
Note The data come from the Ministry of Emergency Management, spaces in blank denote no such losses by floods, and the direct economic losses in some areas do not fully cover the direct economic losses by water engineering projects.

2.4 Disasters and Losses

2.4.1 Summary

In 2023, approximately 52.7893 million person-times were affected by floods, down by 24.7% from the preceding decadal average; 309 people died or went missing, down by 41.6% from the preceding decadal average; 130,000 dwellings collapsed, down by 34.1% from the preceding decadal average; the affected cropland area was 4,633,290 ha, down by 30.8% from the preceding decadal average; the failed cropland area was 559,010 ha. The direct economic loss was 244.575 billion RMB, up by 10.5% from the preceding decadal average; the loss accounted for 0.19% of the annual GDP, down by 29.6% from the preceding decadal average. The four provinces/municipalities of Beijing, Hebei, Heilongjiang, and Henan were the hardest hit; their direct economic losses attributable to floods billed 186.472 billion RMB and accounted for 76.2% of the national total.

图 2-6　2023 年全国洪涝灾害分布图
Figure 2-6　Overview of flood disasters in China in 2023

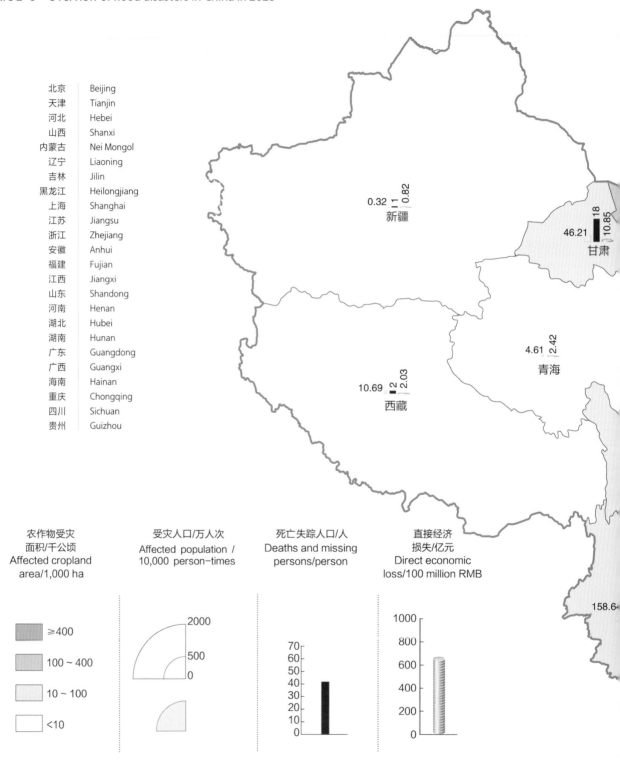

2 洪涝灾害防御

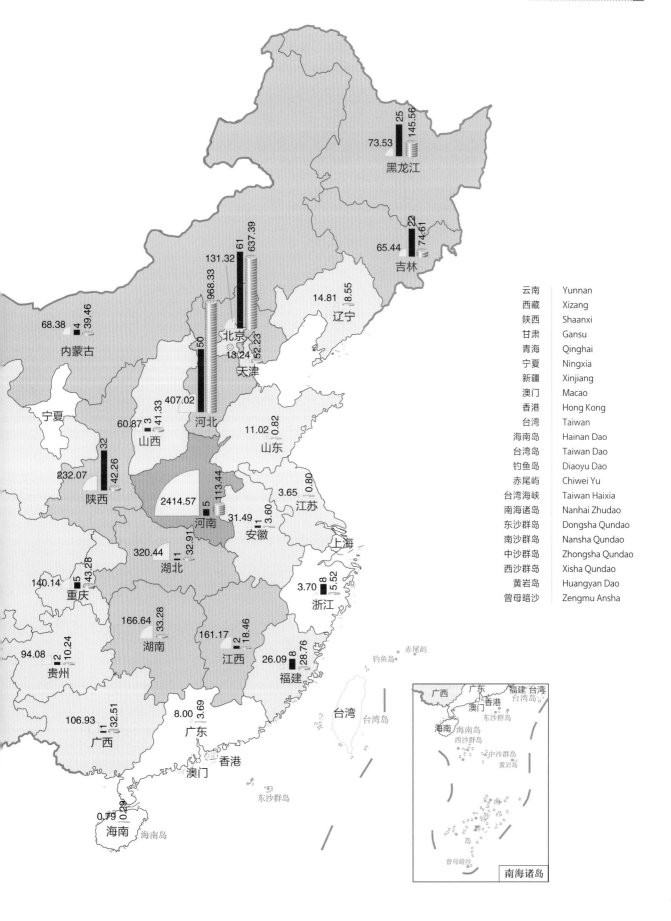

表 2-2 2023 年全国因洪涝农作物受灾面积、农作物绝收面积、倒塌房屋情况

Table 2-2 Cropland affected and failed, collapsed dwellings by floods in China in 2023

地区 Province	农作物受灾面积 / 千公顷 Affected cropland area / 1,000 ha	农作物绝收面积 / 千公顷 Failed cropland area / 1,000 ha	倒塌房屋 / 万间 Collapsed dwellings / 10,000 rooms
全国 Nationwide	4633.29	559.01	13.00
北京 Beijing	14.47	7.23	1.60
天津 Tianjin	26.24	10.37	0.03
河北 Hebei	375.94	182.90	8.73
山西 Shanxi	67.03	4.50	0.05
内蒙古 Nei Mongol	389.51	71.74	
辽宁 Liaoning	32.53	3.63	0.01
吉林 Jilin	227.81	12.63	0.27
黑龙江 Heilongjiang	396.09	166.89	1.51
上海 Shanghai			
江苏 Jiangsu	23.84	3.55	
浙江 Zhejiang	2.85	0.37	
安徽 Anhui	27.27	0.57	
福建 Fujian	18.70	1.80	0.01
江西 Jiangxi	176.85	5.80	0.01
山东 Shandong	10.73	0.31	
河南 Henan	1943.76	11.77	0.01
湖北 Hubei	288.27	14.05	0.04
湖南 Hunan	103.53	10.63	0.11
广东 Guangdong	5.60	0.37	0.02
广西 Guangxi	40.80	2.19	0.11
海南 Hainan	0.75	0.10	
重庆 Chongqing	81.03	10.18	0.18
四川 Sichuan	96.31	7.91	0.15
贵州 Guizhou	39.19	4.55	0.02
云南 Yunnan	74.79	10.15	0.01
西藏 Xizang	3.87	0.50	0.04
陕西 Shaanxi	134.44	13.39	0.06
甘肃 Gansu	26.15	0.73	0.03
青海 Qinghai	3.21	0.06	
宁夏 Ningxia			
新疆 Xinjiang	1.73	0.14	

注 数据来源于应急管理部，空白表示无灾情。

Note The data come from the Ministry of Emergency Management, and spaces in blank denote no losses or damages.

2 洪涝灾害防御

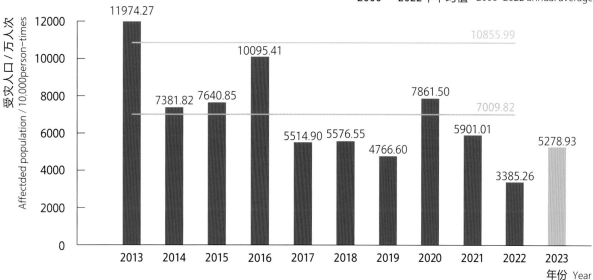

图 2-7　2013—2023 年全国因洪涝受灾人口
Figure 2-7　Population affected by floods in China during 2013-2023

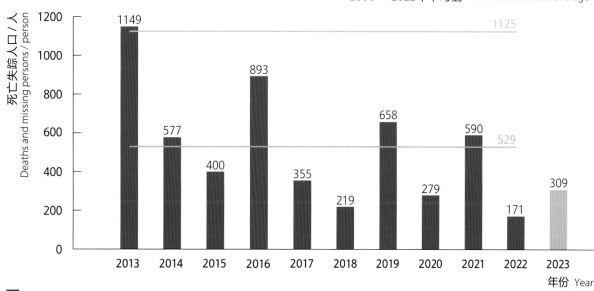

图 2-8　2013—2023 年全国因洪涝死亡失踪人口
Figure 2-8　Deaths and missing persons attributed to floods in China during 2013-2023

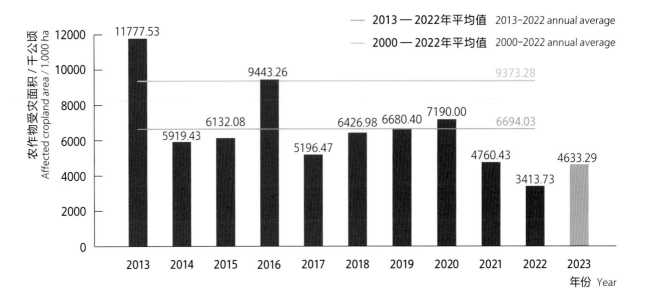

注 2019—2023 年数据来源于应急管理部。
Note The data during 2019-2023 come from the Ministry of Emergency Management.

图 2-9 2013—2023 年全国因洪涝农作物受灾面积
Figure 2-9 Cropland area affected by floods in China during 2013-2023

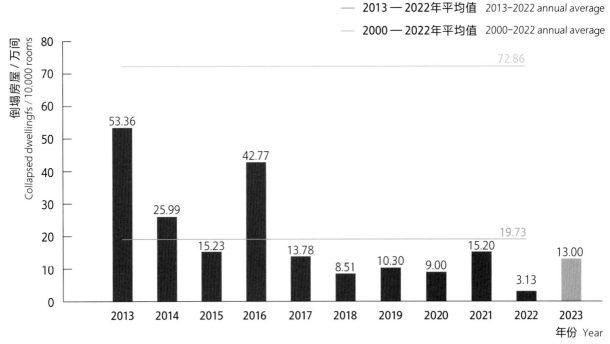

注 2019—2023 年数据来源于应急管理部。
Note The data during 2019-2023 come from the Ministry of Emergency Management.

图 2-10 2013—2023 年全国因洪涝倒塌房屋
Figure 2-10 Collapsed dwellings attributed to floods in China during 2013-2023

2 洪涝灾害防御

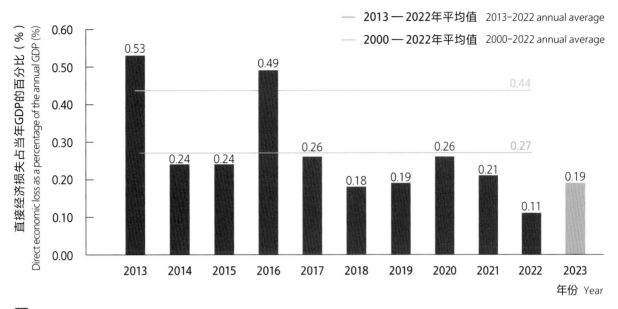

注 2019—2023 年数据来源于应急管理部。
Note The data during 2019-2023 come from the Ministry of Emergency Management.

图 2-11 2013—2023 年全国因洪涝直接经济损失占当年 GDP 的百分比
Figure 2-11 National direct economic losses attributed to floods as a percentage of the annual GDP in China during 2013-2023

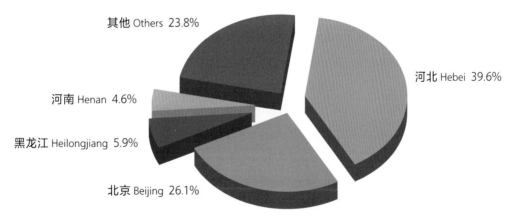

注 数据来源于应急管理部。
Note The data come from the Ministry of Emergency Management.

图 2-12 2023 年全国因洪涝直接经济损失分布
Figure 2-12 A regional break-down of direct economic losses attributed to floods in 2023

2.4.2 水利工程设施灾损情况

2023 年，全国有 30 个省（自治区、直辖市）的水利工程设施因洪涝发生损坏，共造成 1507 座水库（其中 25 座大型、132 座中型、1350 座小型）、29013 处 5607.85 千米堤防、30653 处护岸、3610 座水闸、16767 座塘坝、49325 处灌溉设施、3852 个水文测站、17410 眼机电井、2503 座机电泵站、507 座水电站（其中 5 座大中型、502 座小型）不同程度受损，水利工程设施直接经济损失 633.66 亿元，较前 10 年平均值增加 54%。其中，水利部直属工程中有 1 座大型水库、704 处 15.07 千米堤防、371 处护岸、3 座水闸、64 个水文测站不同程度受损。

2.4.2 Losses and damages to water projects and facilities

In 2023, water projects and facilities in 30 provinces/autonomous regions/municipalities were damaged due to flooding. In total, 1,507 reservoirs (25 large, 132 medium-sized and 1,350 small), 29,013 embankments with a total length of 5,607.85 km, 30,653 bank revetments, 3,610 sluices, 16,767 small pond reservoirs, 49,325 irrigation facilities, 3,852 hydrologic stations, 17,410 electromechanical wells, 2,503 electromechanical pumping stations, and 507 hydropower stations (5 large and medium-sized and 502 small) were damaged to varying degrees. The direct economic loss billed 63.366 billion RMB, up by 54% from the preceding decadal average. Among the projects directly managed by MWR, 1 large reservoir, 704 embankments totaling 15.07 km in length, 371 bank revetments, 3 sluices and 64 hydrologic stations were damaged to varying degrees.

表 2-3 2023 年全国水利工程设施灾损情况

Table 2-3　Losses and damages to water projects and facilities in 2023

地区 Province	损坏水库 / 座 Damaged reservoirs/ number		损坏堤防 Damaged dikes		损坏 护岸 / 处 Damaged revetments/ number	损坏 水闸 / 座 Damaged sluices/ number	损坏 塘坝 / 座 Damaged small pond reservoirs/ number	损坏水文测站 / 个 Damaged hydrologic stations/ number	损坏 水电站 / 座 Damaged hydropower stations/ number	水利工程设施直接经济 损失 / 亿元 Direct economic loss by water projects and facilities/100 million RMB
	大中型 Large and medium- size	小型 Small	数量 / 处 Number of sites	长度 / 千米 Length/km						
全国 Nationwide	157	1350	29013	5607.85	30653	3610	16767	3852	507	633.66
北京 Beijing	9	23	265	188.38	441	13	74	316	2	189.38
天津 Tianjin			75	222.97	26	201		14		5.13
河北 Hebei	21	80	6223	1488.31	6233	243	466	587	79	164.54
山西 Shanxi	4	27	916	186.00	144	2	2	36	6	6.18
内蒙古 Nei Mongol	6	32	3899	233.96	493	47	23			6.73
辽宁 Liaoning	3	8	970	149.64	1235	13	41	56	35	7.46
吉林 Jilin	14	20	303	656.58	92	56		88		18.84
黑龙江 Heilongjiang	7	55	150	88.52	532	47	208	12	17	14.91
上海 Shanghai										
江苏 Jiangsu		3	3	0.32	4	9				0.14
浙江 Zhejiang			1003	283.28	378	55	123	91	13	3.40
安徽 Anhui		5	342	45.54	867	90	1098	1001	2	4.54
福建 Fujian	7	432	2621	193.08	4664	384	475	282	158	43.64

续表 Continued

地区 Province	损坏水库／座 Damaged reservoirs/number		损坏堤防 Damaged dikes		损坏护岸／处 Damaged revetments/number	损坏水闸／座 Damaged sluices/number	损坏塘坝／座 Damaged small pond reservoirs/number	损坏水文测站／个 Damaged hydrologic stations/number	损坏水电站／座 Damaged hydropower stations/number	水利工程设施直接经济损失／亿元 Direct economic loss by water projects and facilities/100 million RMB
	大中型 Large and medium-size	小型 Small	数量／处 Number of sites	长度／千米 Length/km						
江西 Jiangxi		23	439	32.96	1846	437	1148	145	5	8.41
山东 Shandong	2		31	0.86	9	14	1			0.31
河南 Henan	16	55	1103	185.63	1241	239	188	8	2	10.16
湖北 Hubei	10	104	965	274.08	3136	590	2618	132	21	22.71
湖南 Hunan			520	48.60	3090	322	1502	38	16	11.39
广东 Guangdong	33	74	1667	299.10	1096	345	317	63	33	29.18
广西 Guangxi	2	37	986	159.24	1601	229	111	135	23	14.91
海南 Hainan	1	13	3	0.72	12	7	6			0.73
重庆 Chongqing	1	143	1340	82.86	901	96	3746	342	61	12.49
四川 Sichuan	14	141	592	124.94	722	88	4417	163	7	20.71
贵州 Guizhou	1	5	308	43.82	93	9	19		1	1.66
云南 Yunnan		14	643	105.71	540	9	39	23	24	5.93
西藏 Xizang			1		310	96.64	8		12	1.42
陕西 Shaanxi	2	36	2332	290.96	545	31	95	196	2	17.25

续表 Continued

地区 Province	损坏水库/座 Damaged reservoirs/number		损坏堤防 Damaged dikes		损坏护岸/处 Damaged revetments/number	损坏水闸/座 Damaged sluices/number	损坏塘坝/座 Damaged small pond reservoirs/number	损坏水文测站/个 Damaged hydrologic stations/number	损坏水电站/座 Damaged hydropower stations/number	水利工程设施直接经济损失/亿元 Direct economic loss by water projects and facilities/100 million RMB
	大中型 Large and medium-size	小型 Small	数量/处 Number of sites	长度/千米 Length/km						
甘肃 Gansu		2	93	51.55	15	9	4	16		5.94
青海 Qinghai	1	1	95	23.12	41	2	2	2		1.89
宁夏 Ningxia	2	14	10	1.25	135		32	18		0.57
新疆 Xinjiang		2	102	34.14	142	20		24		1.35
水利部直属 Directly managed by MWR		1	704	15.07	371	3		64		1.76

注 空白表示无灾情。
Note Spaces in blank denote no such losses or damages.

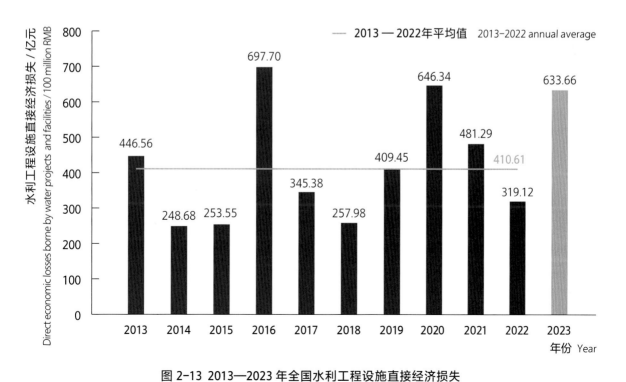

图 2-13 2013—2023 年全国水利工程设施直接经济损失

Figure 2-13 Direct economic losses borne by water projects and facilities in China during 2013-2023

2023年，全国因洪涝造成的水利设施损毁主要集中在北京、河北两省（直辖市），水利工程设施直接经济损失占全国的55.8%。

In 2023, losses and damages to water facilities attributed to floods nationwide were mainly borne by the two provinces/municipalities of Beijing and Hebei, and their direct economic losses accounted for 55.8% of the national total of its kind.

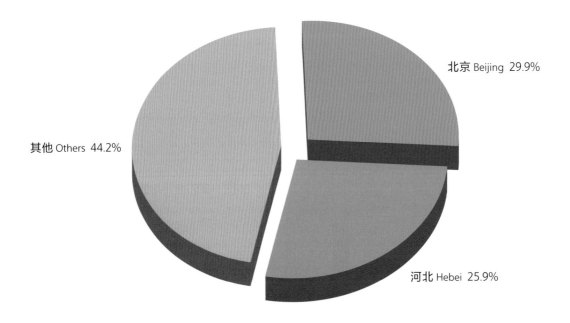

图 2-14 2023 年全国水利工程设施直接经济损失分布

Figure 2-14 A provincial break-down of direct economic losses borne by water projects and facilities in China in 2023

2.5 防御工作

水利部锚定"人员不伤亡、水库不垮坝、重要堤防不决口、重要基础设施不受冲击"目标，贯通雨情、汛情、险情、灾情"四情"防御，强化预报、预警、预演、预案"四预"措施，绷紧降雨—产流—汇流—演进、流域—干流—支流—断面、总量—洪峰—过程—调度、技术—料物—队伍—组织"四个链条"，完善体制、机制、法治、责任制"四制（治）"体系，构建气象卫星和测雨雷达、雨量站、水文站组成的雨水情监测预报"三道防线"，有力有效组织开展洪水灾害防御工作。

2.5 Prevention and Control

The MWR focused relentlessly on achieving its goals of "zero casualties, no dam breaches, no key embankment failures, and no disruption to vital infrastructure". This commitment led to an integrated approach to managing rainfall, water regimes, hazards, and disasters. Central to this strategy were the four preemptive pillars of disaster prevention: forecasting, early warning, exercising, and contingency planning. These efforts entailed strengthening the process management from rainfall, runoff generation, confluence to evolution, from the basin scale, mainstreams, tributaries to sections, from flood volume, flood peak, flood processes to scheduling, and from technology, materials, teams to organization. Additionally, the institutional, mechanistic, legal, and accountability frameworks were also enhanced to refine the management and governance system. A robust defense system was established with three lines of defense for rainfall monitoring and forecasting, comprising meteorological satellites, rain gauge radars, rainfall stations, and hydrological stations, to effectively coordinate flood disaster prevention efforts.

2.5.1 工作部署

汛前，水利部先后召开全国水利工作会议和水旱灾害防御、水库安全度汛、山洪灾害防御等视频会议，逐流域召开防汛抗旱总指挥部工作会议，提早部署防范水旱灾害重大风险；印发《关于做好山洪灾害防御准备工作的通知》，提早部署开展山洪风险隐患排查整治、推进动态预警指标分析应用、修订完善山洪灾害防御预案等重点工作；印发《关于做好 2023 年蓄滞洪区运用准备工作的通知》，安排部署做好蓄滞洪区责任落实、基础数据完善、隐患排查、预案修订、转移避险措施落实等运用准备工作；受中组部委托举办水旱灾害风险管理与水资源刚性约束网上专题培训班，有针对性地调训全国 3400 多名水旱灾害防御行政首长，提高防汛抗旱调度指挥能力；向社会公布 726 座大型水库大坝安全责任人名单。

2.5.1 Work deployment

Prior to the flood season, the MWR convened national conferences on water conservancy and held video conferences on flood and drought disaster prevention, reservoir safety for flood control, and flash flood disaster prevention. Working meetings of Flood Control and Drought Relief Headquarters (hereinafter FDH) were held for each river basin to proactively address major flood and drought risks. Directives such as the *Notice on Preparing for Flash Flood Disaster Prevention* facilitated early deployment of key tasks including hazard identification and rectification, dynamic early warning analysis and application, and revision of disaster prevention plans. Similarly, *The Notice on Preparing for the Utilization of flood detention areas in 2023* facilitated preparatory measures such as ensuring accountability, guaranteeing basic data, identifying and rectifying potential risks, revising plans, and implementing evacuation and relocation measures. A specialized online training program, endorsed by the Organization Department of the CPC Central Committee, was conducted to enhance the flood control and drought relief command capabilities of over 3,400 administrative leaders nationwide. Furthermore, a list of responsible personnel for the safety of 726 major reservoirs was made public.

2.5.2 隐患排查整改

汛前，派出由部领导和总师带队的工作组赴重点流域、重点区域、重点工程，针对水旱灾害防御责任制落实、水情监测预报预警、防洪预案修订与演练、防汛"四预"工作、大中型水库调度运用和小型水库安全度汛、山洪灾害防御和南水北调中线工程安全度汛等开展检查，对发现的问题以"一省一单"督促整改。组织各地采取市（县）级全面自查、省级现场重点抽查和线上检查等方式，全面深入排查自动监测站网、监测预警平台、群测群防等山洪灾害防治存在的风险隐患，建立问题清单、整改清单、责任清单，共整治山洪隐患1.1万处。

2.5.2 Risks identification and rectification

Prior to the flood season, teams led by ministerial officials were dispatched to key river basins, regions, and projects to assess the implementation of flood and drought disaster prevention measures. The inspection focused on the implementation of the accountability system, water regime monitoring, forecasting and early warning, revision and rehearsal of flood control plans, the enhancement of the four preemptive pillars, the operation and dispatching of large and medium-sized reservoirs, the safety of small reservoirs during the flood season, flash flood disaster prevention, and the safety of the central route of the South-to-North Water Diversion Project. Identified issues were promptly addressed through a province-specific approach. Various inspection methods, including comprehensive self-inspections at the city (county) level, targeted on-site inspections at the provincial level, and online inspections, were organized. A thorough investigation was conducted to identify risks and hazards in flash flood disaster prevention and control, such as automatic monitoring station networks, monitoring and early warning platforms, and crowd-sourced disaster prevention efforts. Lists of issues, corrective actions, and responsibilities were established. A total of 11,000 flash flood hazards were rectified.

2.5.3 调度演练

水利部组织开展洪水调度和防御演练，选取历史典型洪涝灾害案例进行模拟实战并检验队伍；督促指导地方以群众自主转移避险、突发灾害应急处置为重点，开展山洪灾害防御实战演练，提升基层干部群众应急避险和自救互救能力。水利部长江水利委员会（以下简称"长江委"）组织开展长江"1999+"洪水防洪调度演练和澧水流域江垭、皂市水库防汛抢险演练。水利部黄河水利委员会（以下简称"黄委"）组织开展黄河防御大洪水调度演练，重点演练雨水情监测预报、洪水防御水工程调度决策、防洪工作部署、应急抢险处置和后勤保障等环节。水利部淮河水利委员会（以下简称"淮委"）会同河南、安徽、江苏等省水利厅，将郑州"7·20"特大暴雨雨型移置到淮河上游桐柏山区，开展淮河防洪调度"四预"演练；沂沭泗水利管理局将郑州"7·20"特大暴雨雨型移置到沂沭泗河水系南四湖片，开展南四湖防洪"四预"推演。水利部海河水利委员会（以下简称"海委"）组织河北、山西、河南、山东等省水利厅及海委漳卫南运河管理局开展漳卫河洪水防御联合演练，模拟漳河发生重现期超50年洪水叠加卫河发生重现期超30年洪水，检验流域统一调度和协调联动机制。水利部珠江水利委员会（以下简称"珠江委"）以"68·6"全流域型洪水为背景，联合广西、广东等省（自治区）有关部门开展流域防洪调度及超标准洪水应急监测演练。水利部松辽水利委员会（以下简称"松辽委"）组织开展松花江流域典型洪水防洪调度演练，选取1998年和2013年典型洪水，采用松辽流域水旱灾害防御系统和全景数字嫩江平台初步建设成果分析研判流域雨水情和防汛形势。水利部太湖流域管理局（以下简称"太湖局"）组织浙江省水利厅及嘉兴市、嘉善县水利局开展太湖流域防洪调度演练，检验应对大洪水的实战能力。

2.5.3 Dispatching and exercising

The MWR organized flood dispatch and defense drills, simulating typical historical flood cases for practical validation. Local authorities were guided to focus on self-organized evacuation by the public and emergency response to sudden disasters, and to conduct practical exercises to enhance the emergency evacuation and self-help/mutual assistance capabilities of rank-and-file officials and the public. The Changjiang Water Resources

Commission (Changjiang Commission hereinafter) organized "1999+" flood control and dispatch drills for the Yangtze River, and flood control and rescue drills for the Jiangya and Zaoshi reservoirs in the Li River Basin. The Yellow River Conservancy Commission (Yellow River Commission hereinafter) organized drills for defending against major floods on the Yellow River, focusing on simulating rainfall monitoring and forecasting, decision-making for dispatching water conservancy projects for flood control, flood control work deployment, emergency rescue and disposal, and logistical support. The Huaihe River Commission (Huaihe Commission hereinafter), together with the water resources departments of Henan, Anhui, and Jiangsu Provinces, relocated and simulated the rainstorm pattern of the "July 20" extreme rainstorm disaster in Zhengzhou to the Tongbai Mountain area in the upper reaches of the Huaihe River, conducting drills for flood dispatch and the four preemptive pillars of forecasting, early warning, exercising, and contingency planning; the Yi-Shu-Si River Basin Administration simulated and relocated the rainstorm pattern of the "July 20" extreme rainstorm disaster in Zhengzhou to the four lakes of Weishan, Zhaoyang, Dushan, and Nanyang, conducting drills of forecasting, early warning, exercising, and contingency planning for flood prevention. The Haihe River Water Conservancy Commission (Haihe Commission hereinafter) organized joint flood defense drills for the rivers of Zhanghe and Weihe in Hebei, Shanxi, Henan, and Shandong Provinces, simulating a flood event with a recurrence period of over 50 years in the Zhanghe River combined with a flood event with a recurrence period of over 30 years in the Weihe River, testing the unified and coordinated dispatch mechanisms in the basin. The Pearl River Water Resources Commission (Pearl River Commission hereinafter), with the relevant departments of Guangxi and Guangdong Provinces/Autonomous Regions, conducted basin-wide flood dispatch and emergency response and monitoring drills for extreme floods, simulating the basin-wide flood in June, 1968. The Songliao River Water Resources Commission (Songliao Commission hereinafter) organized flood control dispatch drills for typical floods in the Songhua River Basin, selecting and simulating typical floods from 1998 and 2013, and analyzing and judging the basin's rainfall and water regime with the assistance of the Songliao River Basin's flood and drought prevention system and the comprehensive digital platform of the Nenjiang River. The Taihu Basin Authority (Taihu Authority hereinafter), together with the water resources department of Zhejiang Province and the water authorities of Jiaxing City and Jiashan County, organized flood dispatch drills for the Taihu Basin, testing the coping capacities against major floods.

2.5.4 监测预报预警

水利部强化"四预"措施，牢牢掌握防御工作主动权，滚动发布未来 2 小时降水短临预警，并"点对点"发至地方；逐日以"一省一单"方式将山洪灾害风险区域及点位发至地方，累计靶向预警 1057 省次，提醒做好局地强降水防范应对；持续开展山洪灾害监测预警抽查，累计抽查县级山洪灾害监测预警情况 1592 县次；会同中国气象局制作发布山洪灾害气象预警 139 期（其中中央电视台播出 34 期）。全国水利部门滚动发布 1452 条河流 3141 个断面洪水预报 36.61 万站次，向防汛责任人和社会公众发布江河洪水干旱水情预警 1558 次，编制海河、松花江流域洪水演进动态 61 期；利用山洪灾害监测预警平台向 989 万名防汛责任人发送预警短信 4679.9 万条，启动预警广播 34.7 万次，依托"三大运营商"向社会公众发布预警短信 20.4 亿条。

长江委密切监视雨水情变化，及时发布各类雨水情信息，通过汛情通报对强降水落区进行"叫应"提醒，协调长江流域气象中心新增共享 60 部单站雷达基数据和 80 部雷达组合的流域反射率拼图网格产品，完善长江流域雨水情监测预报"三道防线"。黄委密切监视流域天气形势变化，降水及洪水演进期间，合理布置测次，周密组织、精细测报，通过传真、短信、微信、邮件等多种途径向各级防汛单位及时发送雨水情实况、预警预报及相关分析成果。淮委针对每次强降水过程，及时预测降水分布、降水总量，提升"四预"能力，深化流域气象水文交流合作，积极构筑雨水情监测预报"三道防线"。海委坚持关口前移、"预"字当先，持续强化"四预"措施，构建纵向到底、横向到边的水旱灾害防御矩阵，强化气象水文技术融合，逐河、逐库、逐站开展精细化滚动预报；强化"以测补报"，洪水演进过程中派出 6 支水文应急监测队，分赴重点河段和关键部位，与京津冀水文部门联合开展应急监测。珠江委运用初步建成的水旱灾害防御"四预"平台，逐河、逐站、逐库滚动开展短期雨水情预测预报，关键期洪水预报精度达 90%；按照"洪水预报滚动通报＋短临降雨点对点预警"机制，及时将风险预警发送到防御一线。松辽委进一步加强与流域气象中心及流域各省（自治区）水文部门沟通协作，开展松辽流域专用洪水预报模型研究，不断提高预报精度，延长洪水预见期，关键场次洪水预报精度达到 85% 以上，重点地区达到 90% 以上。太湖局密切跟踪水雨情变化，每日与气象部门会商，滚动开展太湖及河网代表站水位预测预报，台风"杜苏芮"影响期间，每日 2 次与华东区域气象中心会商。

2.5.4 Monitoring, forecasting and early warning

The MWR strengthened forecasting, early warning, exercising and contingency planning to firmly grasp the initiative in flood defense. Short-term precipitation warnings for the next two hours were continuously issued and sent directly to local authorities on a point-to-point basis. Every day, areas and sites at risk of flash flood disasters were reported to local authorities on a province-specific basis to remind them of preparing for localized heavy rainfall, accumulating 1,057 province-times of targeted warnings. Continuous inspections of flash flood disaster monitoring and early warning systems were conducted, with a total of 1,592 county-level inspections carried out. In collaboration with the China Meteorological Administration, the Ministry produced and released 139 editions of meteorological warnings for flash flood disasters, 34 of which were broadcast by China Central Television. Water authorities nationwide released in a non-stop manner 366,100 station-times of flood forecast for 3,141 cross sections of 1,452 rivers, and issued flood and drought warning 1,558 times to flood control authorities and the general public. A total of 61 updates were published on the flood evolution in the Haihe and Songhua River basins. The monitoring and early warning platform for flash flood disasters sent 46.799 million text messages to 9.89 million flood control personnel, made 347,000 broadcasts, and sent 2.04 billion warning text messages to the public through the three major telecom operators.

The Changjiang Commission closely monitored changes in rainfall, timely released various rainfall information, and issued "call and response" reminders to areas with heavy rainfall through flood regime notifications. It coordinated with the Yangtze River basin Meteorological Center to add the basin reflectivity mosaic grid products that shared data from 60 single-station radars and 80 radar combinations, enhancing the three-tier defense of rainfall and water regime monitoring and forecasting in the Yangtze River basin. The Yellow River Commission closely monitored changes in regional weather conditions, arranged measurements reasonably during precipitation and flood evolution, and provided detailed and precise forecasts. It timely sent rainfall conditions, early warnings, and related analysis results to various flood control authorities through means such as fax, SMS, WeChat, email, etc. The Huaihe Commission effectively responded to strong precipitation events by predicting precipitation distribution and total amounts in a timely manner, enhancing the capabilities of forecasting, early warning, exercising and contingency planning. It

表 2-4 2023 年汛期各流域管理机构监测预报预警情况
Table 2-4 Monitoring, forecasts and early warnings by each river basin commission/authority during the 2023 flood season

流域管理机构 Commission/authority	降水预报/期 Precipitation forecasts/time	洪水预报/期 Flood forecasts/time	洪水预警/站次 Flood warnings/Station-time	预警短信/万条 Alert messages/10,000 pieces
长江委 Changjiang	380	14	22	5.5
黄委 Yellow River	165	244		3.3
淮委 Huaihe	160	82	4	1.22
海委 Haihe	60	41	4	3.6
珠江委 Pearl River		154	2	5.7
松辽委 Songliao	18	70	16	1.46
太湖局 Taihu		196		1.42

注 空白表示无。
Note Blank indicates none.

表 2-5 2023 年国家蓄滞洪区启用情况
Table 2-5 The utilization of national flood detention areas in 2023

流域 River basin	所属水系 River system	蓄滞洪区 Flood detention areas	启用时间 Utilization start time	蓄滞洪量/亿立方米 Retained floodwater/100 million m³
海河流域 Haihe River basin	子牙河水系 Ziya River System	大陆泽 Daluze 宁晋泊 Ningjinpo	7月30日20时 20:00, July 30	6.06
		献县泛区 Xian County Floodplain	8月1日11时 11:00, August 1	1.21
	大清河水系 Daqing River System	小清河分洪区 Xiaoqing River Flood Diversion Area	7月31日12时 12:00, July 31	5.14
		兰沟洼 Langouwa	7月31日23时30分 23:30, July 31	2.29
		东淀 Dongdian	8月1日2时 2:00, August 1	7.87
	漳卫河水系 Zhanghe and Weihe River System	共渠西 Gongquxi	8月1日15时 15:00, August 1	0.14
	永定河水系 Yongding River System	永定泛区 Yongding River Floodplain	8月2日6时 6:00, August 2	2.56

注 空白表示无。
Note Blank indicates none.

deepened cooperation in meteorological and hydrological information exchange within the basin, actively building three lines of defense for rainfall and water regime monitoring and forecasting. The Haihe Commission, prioritizing forecasting, continuously strengthened the four preemptive pillars, constructed a vertical and horizontal water and drought disaster defense matrix, enhanced the integration of meteorological and hydrological technologies, and conducted refined rolling forecasts river by river, reservoir by reservoir, and station by station. It supplemented reporting with prediction, and sent six hydrological emergency monitoring teams during the flood process, deploying them to key river sections and critical locations, and conducting emergency monitoring in conjunction with the hydrological departments of Beijing, Tianjin, and Hebei. The Pearl River Commission, tapping the platform of the four preemptive pillars, conducted non-stop short-term rainfall and flood forecasts river by river, station by station, and reservoir by reservoir. The accuracy of flood forecasts during critical periods reached 90%. Risk warnings were sent to the front line of flood control in a timely manner under the mechanism of "non-stop flood forecast notifications + point-to-point short-term rainfall warnings". The Songliao Commission further strengthened communication and cooperation with meteorological centers within the basin and provincial hydrological departments. It conducted research on specialized flood forecasting models for the Songliao River Basin, continuously improving forecast accuracy and extending flood forecasting lead times, with the forecast accuracy exceeding 85% for major flood events and 90% for key regions. The Taihu Authority closely monitored changes in water and rainfall conditions, consulted daily with meteorological departments, and released water level forecasts for Taihu Lake and representative stations of the surrounding river network in a non-stop manner. Facing the impact of Typhoon Doksuri, it consulted with the Regional Meteorological Center of Eastern China twice every day.

2.5.5　水工程调度

水利部及各流域管理机构、地方各级水利部门坚持以流域为单元，联合调度运用流域防洪工程体系，综合考虑干支流、上下游、左右岸，采取"拦、分、蓄、滞、排"措施，统筹安排洪水出路，下达调度指令 2.05 万道，调度运用 4512 座次大中型水库、拦蓄洪水 603 亿立方米，启用 8 处国家蓄滞洪区、蓄洪滞洪 25.3 亿立方米，有效减轻了下游地区防洪压力。

长江委统筹考虑供水、发电、航运等多方用水需求，汛期调度三峡水库 6 次拦蓄 20000 立方米每秒以上的中小洪水过程，平稳调控三峡水库下泄流量；防御汉江秋汛期间，先后发出 17 道调度令精细调度丹江口水库，会同陕西、湖北、河南省水利厅科学联合调度石泉、安康、潘口、三里坪和鸭河口等干支流控制性水库拦洪削峰错峰，2 次洪水过程累计拦蓄洪水 17.5 亿立方米。黄委针对泾渭河、汾河、伊洛河和大汶河较大洪水过程，组建联合工作专班，滚动调算黄河下游干支流水库群联合调度方案，统筹水库预泄和实时调度应对措施。淮委有效应对流域强降水过程，协调河南省调度淮河干流出山店、洪汝河板桥、薄山、宿鸭湖等大型水库提前预泄、拦洪削峰，拦蓄洪水约 15.3 亿立方米；协调安徽省提前调度临淮岗、蚌埠闸预排河道底水，降低河道水位，避免了淮河王家坝站出现超警戒洪水；协调江苏省加强洪泽湖调度，共排泄洪水 140 亿立方米。海委强化大中型水库统一调度，严格汛限水位管控，督导各地调度水库

2.5.5 Water project scheduling

The MWR, along with river basin commissions/authorities and local water authorities, adhered to the principle of watershed management. They coordinated and utilized the watershed flood control engineering system, considering both main and tributary rivers, upstream and downstream areas, as well as left and right banks. Measures such as blocking, diversion, storage, detention and retention, and discharge were employed. A total of 20,500 dispatch orders were issued, managing the operation of 4,512 large and medium-sized reservoirs, retaining and storing 60.3 billion m^3 of floodwater. Additionally, 8 national flood detention areas were activated to hold 2.53 billion m^3 of floodwater, effectively alleviating flood control pressure downstream.

The Changjiang Commission, taking into account various demands such as water supply, power generation, and navigation, coordinated the flood season operation of the Three Gorges Reservoir, intercepting six instances of medium and small floods exceeding 20,000 m^3/s, stabilizing the discharge flow of the Reservoir. During the autumn flood season of the Hanjiang River, 17 dispatch orders were successively issued to finely adjust the operation of the Danjiangkou Reservoir. In conjunction with the water resources departments of Shaanxi, Hubei, and Henan provinces, the Commission utilized controlling reservoirs, such as Shiquan, Ankang, Pankou, Sanliping, and Yahekou, along the main and tributary streams of the Yangtze River to intercept floods, cut and delay flood peaks. In this way, a total of 1.75 billion m^3 of floodwater was retained during the two flood events.

The Yellow River Commission, in response to major flood events in the Jinghe River, Weihe River, Fenhe River, Yiluo River, and Dawen River, established joint task forces to develop coordinated operation plans for reservoir groups downstream of the Yellow River, and coordinate reservoir pre-discharge and real-time dispatch measures.

The Huaihe Commission effectively responded to heavy rainfall events in the basin. In Henan Province, it coordinated pre-discharge and interception of floods at major reservoirs such as Shandian on the mainstream Huaihe, Banqiao, Boshan, and Suyahu on the Hongru River to cut flood peaks, retaining approximately 1.53 billion m^3 of floodwater. In Anhui Province, its coordination efforts included drawing down instream water levels in advance at Linhuaigang and Bengbu water gates to prevent floods above the warning level from happening at the Wangjiaba Station on the Huaihe River. In Jiangsu Province, efforts were made to strengthen the coordinated dispatch of the Hongze Lake, with a total of 14 billion m^3 of floodwater being discharged.

预泄腾库迎洪；洪水来临时，组织大中型水库充分拦洪，累计拦蓄洪水 28.5 亿立方米；强化河道闸坝科学调度，组织实施河网预泄，橡胶坝全部塌坝运行，调度北运河北关、土门楼枢纽，运用运潮减河、青龙湾减河向潮白河分泄洪水，调度委属屈家店枢纽、独流减河进洪闸与防潮闸最大限度宣泄洪水；强化蓄滞洪区安全运用，及时督导提醒有关地方提前做好群众避险转移准备。珠江委在"龙舟水"期间，调度贺江上游龟石、合面狮等水库拦蓄洪水 1.28 亿立方米，指导下游都平等水库预泄；台风"海葵"影响期间，调度棉花滩、益塘、合水、长潭等上游水库精准拦蓄洪水 1.75 亿立方米，调度高陂等水库预泄腾库 0.85 亿立方米。松辽委按照雨前预泄迎汛、雨间拦洪错峰的原则，科学调度察尔森、尼尔基、丰满、白山 4 座直调水库 13 次，累计拦蓄洪水 17.76 亿立方米；应对嫩江洮儿河洪水过程中，精细调度察尔森水库 38 小时零出流，有效削减洪峰，减轻下游防洪压力。太湖局针对梅雨期流域 4 轮强降水，提前调度望虞河工程、太浦河工程及新孟河工程加大排水，并督促流域沿长江、杭州湾口门加大排水力度，降低太湖、京杭运河及河网水位，强降水期间压减太浦闸排水流量，为下游地区让出排涝通道。

The Haihe Commission strengthened the unified operation of large and medium-sized reservoirs, enhanced the supervision of flood control level compliance, and supervised and guided the pre-discharge of reservoirs to make room for incoming floods. When floods arrived, large and medium-sized reservoirs were organized to intercept floods fully, holding a total of 2.85 billion m³ of floodwater; scientific operation of water gates and dams was intensified, with pre-discharge operations implemented in river networks, all rubber dams were deflated, the Beiguan and Tumenlou water gates on Beiyun River were dispatched, measures were taken to divert floodwaters through the flood release channels of Yunchao and Qinglongwan to the Chaobai River, and the water gates of Qujiadian and Duliu Flood Release Channel were employed to ensure flood discharge at the maximum level; efforts were also made to ensure the safe utilization of the flood detention areas and the timely evacuation and relocation of local residents.

During the Dragon Boat rainy season from late May till late June, the Pearl River Commission, by coordinating reservoirs in the upper reaches of the Hejiang River such as Guishi and Hemianshi, intercepted 128 million m³ of floodwater and guided downstream reservoirs (such as Duping) to release water in advance. Facing the impact of Typhoon Haikui, the Commission retained 175 million m³ of floodwater with the assistance from upstream reservoirs including Mianhuatan, Yitang, Heshui, and Changtan, and made available 85 million m³ of floodwater storage capacity through pre-discharge at reservoirs like Gaobei.

Adhering to the principle of pre-releasing water before rainfall and intercepting floods to cut peaks during rainfall, the Songliao Commission scientifically managed reservoirs such as Chaersen, Nierji, Fengman, and Baishan, retaining and holding a total of 1.776 billion m³ of floodwater. During the flood of the Nenjiang River and Tuo'er River, Chaersen Reservoir was finely operated with zero outflow for 38 hours, effectively reducing the flood peak and alleviating flood control pressure downstream.

The Taihu Authority responded to four times of heavy rainfall during the plum rain season in the basin by advancing the operation of the Wangyu River, Taipu River, and Xinmeng River projects to increase drainage capacity. It also urged increased drainage at water gates along the Yangtze River and the Hangzhou Bay, reducing water levels in Taihu Lake, the Beijing-Hangzhou Grand Canal, and river networks. During periods of heavy rainfall, the discharge flow at the Taipu Watergate was reduced to provide drainage pathways for downstream areas.

表 2-6 2023 年水利部本级洪水防御应急响应启动情况
Table 2-6 Initiation of flood disaster prevention emergency response by the Ministry of Water Resources in 2023

序号 No.	启动日期 Start date	响应级别 Response level	针对区域 Targeted area	终止日期 End date
1	6月19日 June 19	IV	江苏、浙江、安徽、福建、江西、湖北、湖南、广西、贵州、云南 Jiangsu, Zhejiang, Anhui, Fujian, Jiangxi, Hubei, Hunan, Guangxi, Guizhou, Yunnan	6月26日 June 26
2	7月4日 July 4	IV	内蒙古、辽宁、吉林、黑龙江 Nei Mongol, Liaoning, Jilin, Heilongjiang	7月10日 July 10
3	7月6日 July 6	IV	江苏、安徽、湖北、重庆、贵州 Jiangsu, Anhui, Hubei, Chongqing, Guizhou	安徽、湖北、重庆、贵州7月10日，江苏7月15日 Anhui, Hubei, Chongqing, Guizhou July 10, Jiangsu July 15
4	7月10日 July 10	IV	山东、四川 Shandong, Sichuan	7月15日 July 15
5	7月12日 July 12	IV	甘肃、青海 Gansu, Qinghai	7月15日 July 15
6	7月15日 July 15	IV	福建、广东、广西、海南、贵州、云南 Fujian, Guangdong, Guangxi, Hainan, Guizhou, Yunnan	7月20日 July 20
7	7月19日 July 19	IV	北京、天津、河北 Beijing, Tianjin, Hebei	7月23日 July 23
8	7月25日 July 25	IV	浙江、安徽、福建、江西、河南、湖北、广东 Zhejiang, Anhui, Fujian, Jiangxi, Henan, Hubei, Guangdong	浙江、安徽、江西、湖北7月30日 Zhejiang, Anhui, Jiangxi, Hubei July 30
9	7月27日 July 27	III	福建、广东 Fujian, Guangdong	7月30日 July 30
10	7月28日 July 28	III	北京、天津、河北、山西、山东、河南 Beijing, Tianjin, Hebei, Shanxi, Shandong, Henan	山西、山东、河南8月6日 Shanxi, Shandong, Henan August 6
11	7月30日 July 30	II	北京、天津、河北 Beijing, Tianjin, Hebei	北京8月15日，天津、河北8月31日 Beijing August 15, Tianjin, Hebei August 31
12	8月2日 August 2	IV	内蒙古、辽宁、吉林、黑龙江 Nei Mongol, Liaoning, Jilin, Heilongjiang	辽宁8月15日 Liaoning August 15
13	8月6日 August 6	III	内蒙古、吉林、黑龙江 Nei Mongol, Jilin, Heilongjiang	内蒙古8月14日，吉林、黑龙江8月15日 Nei Mongol August 14, Jilin, Heilongjiang August 15
14	8月25日 August 25	IV	江苏、安徽、山东、河南、湖北、湖南、重庆、四川、贵州、云南、西藏、陕西、甘肃 Jiangsu, Anhui, Shandong, Henan, Hubei, Hunan, Chongqing, Sichuan, Guizhou, Yunnan, Xizang, Shaanxi, Gansu	8月28日 August 28

续表 Continued

序号 No.	启动日期 Start date	响应级别 Response level	针对区域 Targeted area	终止日期 End date
15	8月26日 August 26	IV	山西 Shanxi	8月28日 August 28
16	8月30日 August 30	IV	福建、江西、湖南、广东、广西、海南 Fujian, Jiangxi, Hunan, Guangdong, Guangxi, Hainan	福建、江西、湖南、海南9月8日，广东、广西9月16日 Fujian, Jiangxi, Hunan, Hainan September 8, Guangdong, Guangxi September 16
17	9月18日 September 18	IV	重庆、四川、陕西 Chongqing, Sichuan, Shaanxi	10月7日 October 7
18	9月21日 September 21	IV	江苏、安徽、河南 Jiangsu, Anhui, Henan	9月24日 September 24
19	9月30日 September 30	IV	湖北 Hubei	10月7日 October 7
20	10月5日 October 5	IV	福建、广东 Fujian, Guangdong	福建10月9日，广东10月11日 Fujian October 9, Guangdong October 11
21	10月18日 October 18	IV	广东、广西、海南 Guangdong, Guangxi, Hainan	10月24日 October 24
22	12月19日 December 19	IV	甘肃、青海、新疆（地震） Gansu, Qinghai, Xinjiang (earthquake)	12月27日 December 27

表2-7 2023年各流域管理机构会商及洪水防御应急响应启动情况

Table 2-7 Consultations held and emergency responses initiated by river basin commissions authorities for flood disaster prevention in 2023

流域管理机构 Commissions / authorities	会商次数 Consultations	启动次数 Emergency responses				累计时间/天 Accumulation time / days
		I级 Level I	II级 Level II	III级 Level III	IV级 Level IV	
长江委 Changjiang	136				8	54
黄委 Yellow River	53				5	22
淮委 Huaihe	73			1	7	19
海委 Haihe	72	1	1	1	1	39
珠江委 Pearl River	127		1	3	6	57
松辽委 Songliao	82		1	1	3	37
太湖局 Taihu	283		1	2	8	33
合计 Total	814	1	4	8	38	261

注 空白表示未启动。

Note Blank means zero initation.

2.5.6 会商响应

水利部建立并实行主汛期部长"周会商＋场次洪水会商"机制，国家防汛抗旱总指挥部副总指挥、水利部部长李国英先后主持会商15次，并率队深入海河一线指导抗洪抢险工作。水利部密切监视雨情水情汛情旱情，逐日跟踪分析，滚动会商研判161次，启动洪水防御应急响应22次，指导流域管理机构启动洪水防御应急响应51次；水利部及各流域管理机构共派出215个次工作组、专家组，协助指导地方开展防御工作。汛期每天以"一省一单"形式将预报降水量超过预警阈值（50毫米或25毫米）的县（市、区）和水库名单发至相关省级水利部门，提醒做好强降水防范。

2.5.7 部门协作

水利部派员参加应急管理部管理洪涝灾害过程会商28次，对洪水趋势作出预测预报，为抗洪抢险提供支撑；派员参加应急部全国自然灾害灾情会商12次、风险形势会商12次，共同核定全国洪旱灾情，推进灾情数据、洪水预警等信息共享。水利部商财政部安排水利救灾资金35.282亿元，支持地方做好安全度汛隐患整治、防洪工程设施水毁修复、白蚁等害堤动物防治和水利工程震损修复等。商财政部及时拨付河北、天津、河南3省（直辖市）蓄滞洪区运用补偿资金110.98亿元，支持受灾群众尽快恢复生产生活。配合国家发展改革委、财政部落实特别国债项目，加快推进水库、堤防、蓄滞洪区等防洪工程建设，构建现代化雨水情监测网和智慧化调度决策系统，进一步完善流域特别是北方地区主要江河流域防洪工程体系。会同中国气象局完善山洪灾害气象预警发布机制。及时向有关部委提供洪涝灾情趋势预测及洪水预警信息，为做好各行业洪涝灾害防御提供支撑。

2.5.6 Consultation and responses

The MWR established and implemented the mechanism of "regular weekly consultation + consultation on specific flood events". Li Guoying, Vice Commander of FDH and Minister of Water Resources, presided over 15 consultations, and led teams to the front-line of the Haihe River to guide flood control and rescue efforts. The Ministry closely monitored rainfall, water levels, floods, and droughts, conducting daily tracking and analysis. It held 161 consultations, initiated 22 emergency responses to flood disasters, and guided the initiating of 51 emergency responses by river basin commissions/authorities to flood disaster. The Ministry and river basin commissions/authorities dispatched a total of 215 work teams and expert groups to assist and guide local prevention efforts. During the flood season, a list of areas and reservoirs forecast to receive precipitation exceeding the warning threshold (50 mm or 25 mm) was sent daily to relevant provincial water authorities, reminding them to strengthen prevention measures against heavy rainfall.

2.5.7 Cross-sectoral collaboration

The MWR participated in 28 consultations on flood disasters organized by the Ministry of Emergency Management, predicting flood trends and providing support for flood control and rescue efforts. Representatives also attended 12 national consultations on natural disasters and 12 risk assessing consultations, jointly assessing the status of floods and droughts nationwide and promoting the sharing of disaster-related data and flood early warning information. The MWR, in collaboration with the Ministry of Finance, allocated 3.5282 billion RMB in disaster relief to support local efforts in rectifying hidden dangers during flood season, repairing flood control projects damaged by floods, controlling pests such as termites that damage embankments, and restoring water conservancy projects damaged by earthquakes. The two Ministries also promptly allocated 11.098 billion RMB in compensation to Hebei, Tianjin, and Henan provinces/municipalities for the use of flood detention areas, supporting disaster-affected communities to quickly resume production and livelihoods. In coordination with the National Development and Reform Commission and the Ministry of Finance, efforts were made to implement special treasury bond projects, with purposes of accelerating the construction of flood control projects such as reservoirs, embankments, and flood detention areas, establishing a modern rainfall and water regime monitoring network and an intelligent decision-making system for scheduling and dispatching, and further improving the flood control system of major river basins, especially in the north of China. Together with the China Meteorological Administration, the Ministry also improved the mechanism for issuing meteorological warnings for flash flood disasters. Moreover, it provided relevant ministries with real-time forecasts and early warning information on flood disasters to support the flood control and prevention efforts across various sectors.

2.5.8 技术支撑

全国水利系统共派出工作组 5.56 万组次 23.86 万人次赶赴前线，协助指导地方做好水旱灾害防御工作；派出专家组 1.44 万组次 6.74 万人次提供防汛抢险技术支持，指导处置水利工程险情 7826 处。在海河"23·7"流域性特大洪水灾害防御中，中国水利水电科学研究院和海委等单位专家根据海河流域将发生特大暴雨的预测信息，迅速分析预测可能发生的洪灾量级，识别水库、河道、堤防、蓄滞洪区等洪水风险隐患和防汛薄弱环节，模拟预测东淀蓄滞洪区分洪情况，提出运用时机、退洪时间、退洪位置等建议，进一步优化蓄滞洪区运用方案。针对可能对雄安新区造成影响的白沟河左堤重大险情，水利部前方工作组指导地方扩大右堤分洪，同时加固堤防、加固外围围堰、封堵公路桥涵以构筑"三道防线"，全力控制风险，确保重点地区防洪安全。

2.5.8 Technical support

Water authorities of all levels sent out 55,600 group-times of 238,600 person-times in working groups to the front line to assist and guide local efforts in flood and drought disaster prevention. Additionally, 14,400 team-times of 67,400 person-times in expert teams provided technical support for flood control and rescue efforts, guiding the handling of 7,826 hazards of water conservancy projects. In coping with the " 23·7 " extreme flood of the Haihe River basin, experts from institutions such as the China Institute of Water Resources and Hydropower Research and the Haihe Commission, by analyzing the forecast of basin-wide heavy rainfall, rapidly predicted the severity of the flood disaster and identified the hidden risks in reservoirs, river channels, embankments, flood detention areas, and weak links in flood control. In addition, they simulated and predicted flood situations in the Dongdian Flood Detention Area, and made suggestions on the timing, duration, and locations for flood discharge, further optimizing the utilization plan for the flood detention and retention basin. In response to the major hazard at the left embankment of the Baigou River which might pose a threat to the Xiong'an New Area, the on-site working group of the MWR guided local authorities to strengthen flood diversion from the right embankment, reinforce embankments, solidify cofferdams, and seal the bridges and culverts of highways, so as to establish "three lines of defense", contain the risks, and ensure the flood control safety of key regions.

2.6 防御成效

2023 年，水利部门有效实施"四预"举措，科学调度运用水库、河道及堤防、蓄滞洪区，及时指导开展险情抢护，成功应对了江河洪水，全国水库无一垮坝，大江大河干流堤防无一决口。全国减淹城镇 1299 座次，减淹耕地 1073.23 千公顷，避免人员转移 721.28 万人次，最大程度保障了人民群众生命财产安全及重要基础设施安全运行。

表 2-8 2023 年各流域减淹城镇、耕地及避免人员转移情况

Table 2-8 Cities / towns and cropland protected from floods, and population avoiding evacuation in 2023

流域 River/lake basin	减淹城镇 / 座次 Cities/towns cumulatively protected from floods	减淹耕地 / 千公顷 Cropland protected from floods/1,000 ha	避免人员转移 / 万人次 Population avoiding evacuation/10,000 person-times
全国 Nationwide	1299	1073.23	721.28
长江流域 The Yangtze River	128	46.71	13.45
黄河流域 The Yellow River	13	10.40	1.34
淮河流域 The Huaihe River	34	48.20	19.85
海河流域 The Haihe River	112	586.67	520.87
珠江流域 The Pearl River	388	73.91	80.15
松辽流域 The Songhua-Liaohe River	211	217.91	57.21
太湖流域 The Taihu Lakes	413	89.43	28.41

2.6 Effectiveness of Flood Disaster Prevention

In 2023, China's water authorities, tapping into the four preemptive pillars, effectively coordinated the utilization of reservoirs, river channels, embankments, and flood detention areas, provided timely guidance for emergency responses and rescue efforts, and hence coped with river floods successfully, with zero incidents of dam collapse nationwide and embankment breach for main rivers. Cumulatively, 1,299 cities and towns and 1,073,230 ha of farmland were protected from flooding, 7.2128 million people were spared from evacuation and relocation, thus ensuring the safety of people's lives and property as well as the smooth operation of critical infrastructure to the greatest extent possible.

案例 1 海河"23·7"流域性特大洪水调度

针对海河流域性特大洪水，李国英部长8天内7次主持专题会商，逐河系超前部署、逐河系主动出击、逐河系科学防控。水利部启动洪水防御Ⅱ级应急响应，海委、京津冀豫水利部门均启动洪水防御Ⅰ级响应并联动落实各项防御措施。水利部提前1周对海河流域暴雨洪水形势作出研判，关键期洪水预报精度达80.6%；首次应用水利部数字孪生平台，提前预判蓄滞洪区启用时间，动态掌握蓄滞洪区水头演进、滞蓄洪量、淹没范围等调度运用状态，构筑蓄滞洪区水动力学模型进行预演，有力支撑蓄滞洪区调度及运用决策。

科学调控子牙河洪水，联合调度滹沱河岗南、黄壁庄水库，岗南水库拦蓄洪水3.4亿立方米，削峰率99.9%，并为岗南至黄壁庄区间冶河来水错峰近20小时；黄壁庄水库充分拦蓄，控制下泄流量远低于石家庄市安全行洪流量。调度滏阳河上游朱庄、临城等水库充分拦蓄洪水。及时启用大陆泽、宁晋泊、献县泛区3处蓄滞洪区，科学调度艾辛庄枢纽、献县枢纽，保证了滏阳新河、子牙新河等骨干河道行洪安全。

精细调控永定河洪水，调度官厅水库关闸拦蓄全部来水；调度支流斋堂水库运用至历史最高水位，为落坡岭火车站滞留旅客及下游低洼地区群众转移赢得时间；调度卢沟桥枢纽精准分洪，将超量洪水分入大宁、稻田、马厂水库，确保了永定河堤防和大兴国际机场安全，并为永定河泛区人员转移争取了时间；及时启用永定河泛区缓洪滞洪，有效减轻天津市城区防洪压力。

Case 1 Flood control and dispatch during the "23·7" basin-wide extreme floods of the Haihe River basin

In response to the basin-wide extreme floods in the Haihe River basin, Minister Li Guoying presided over seven thematic consultations within eight days, deploying proactive and scientific flood control measures across each river system. The MWR initiated a Level II emergency response for flood control, while the Haihe Commission and the water authorities of Beijing, Tianjin, Hebei, and Henan activated a Level I emergency response for flood control and implemented various control measures in coordination. One week prior, the Ministry analyzed the potential rainstorm and floods in the Haihe River basin and achieved an 80.6% accuracy rate in its flood forecasting for the critical period. For the first time, the Ministry's digital twin platform was employed to predict the activation timing of flood detention areas, to dynamically monitor the movement of the flood, the volume of floodwater retained, and inundation extents in these basins, and to construct a hydraulic model for pre-simulation, in an effort to support the decision-making for their dispatch and utilization.

Scientific management of the Ziya River floods involved coordinated dispatch of the Gangnan and Huangbizhuang Reservoirs on the Hutuo River, with Gangnan Reservoir intercepting 340 million m³ of floodwaters, achieving a peak reduction rate of 99.9%. This action postponed the flood peak of the Yehe River between Gangnan and Huangbizhuang for nearly 20 hours. Huangbizhuang Reservoir retained floodwaters sufficiently to keep the discharge volume significantly below the safe limit for Shijiazhuang City. Upstream reservoirs such as Zhuzhuang and Lincheng on the Fuyang River were also fully utilized for flood retention. The timely activation of flood detention areas at Daluze, Ningjinpo, and the floodplain of the Xian County, alongside scientific scheduling at the Aixinzhuang and Xian County control works, ensured the safety of major waterways such as the Fuyangxin River and Ziyaxin River.

To control the Yongding River floods in a precise manner, the Guanting Reservoir stopped water discharge completely to retain all incoming waters; the Zhaitang Reservoir, on a tributary of the Yongding River, was operated to such an extent that it witnessed a record high water level, buying time for the evacuation of stranded passengers at Luopoling Train Station and residents in downstream low-lying areas; precision flood diversion at the Lugouqiao Control Complex directed excess floodwaters into the Daning, Daotian, and Machang Reservoirs, protecting the Yongding River embankments and the Daxing International Airport, and facilitating the relocation of people from flood-prone areas along the River; the timely activation of the flood detention area at the floodplain of the Yongding River effectively alleviated the flood control pressures for Tianjin City.

有效调控大清河洪水，及时启用小清河、兰沟洼、东淀 3 处蓄滞洪区，最大蓄滞洪量 15.3 亿立方米；调度大清河南支上游王快、西大洋等水库群充分拦蓄洪水，削峰率均超过 90%，平稳控制白洋淀水位，避免大清河南北支洪水遭遇；调度枣林庄、新盖房枢纽和独流减河进洪闸等，根据上下游水势，有序行泄洪水，保证了新盖房分洪道、赵王新河、独流减河等骨干河道行洪安全。

系统调控北运河洪水，调度北运河上游十三陵水库充分拦蓄，削峰率 100%；调度密云、怀柔等水库拦蓄洪水，降低潮白河水位，为北运河洪水东排创造条件；精细调度北关、土门楼枢纽，分别通过运潮减河、青龙湾减河向潮白河分泄洪水，在保证北运河行洪安全的同时，避免了天津大黄堡洼蓄滞洪区启用。

在科学调度运用流域防洪工程体系的同时，强化堤防巡查防守，组织 22 万人巡堤查险，及时处置堤防险情 131 处。通过共同努力，最大程度保障了人民生命安全，确保了流域内重要防洪对象安全，各类水库无一垮坝，重要堤防和蓄滞洪区围堤无一决口，蓄滞洪区内撤退转移近百万人无一伤亡，夺取了海河"23·7"流域性特大洪水防汛抗洪斗争的重大成果。

To manage and control the Daqing River floods, the flood detention areas at Xiaoqing River, Langouwa, and Dongdian were activated in a timely manner, achieving a maximum flood retention volume of 1.53 billion m^3; reservoirs like Wangkuai and Xidayang, located on the upper reaches of the southern branch of the Daqing River, fully retained floodwaters (each achieving a peak reduction rate over 90%), staggered the flood peaks of the southern and northern branches, and stabilized the water level of Baiyangdian Lake; scheduled dispatch at Zaolinzhuang and Xingaifang Control Complexes, and the Duliujian River Floodgate, taking into consideration both the upstream and downstream water regime, ensured orderly flood discharge and thus the safety of key waterways like Xingaifang Diversion Channel, Zhaowangxin River, and Duliujian River.

To control the floods in Beiyun River, Shisanling Reservoir on its upper reaches was used to effectively retain floodwater and shaved the flood peak discharge by 100%; reservoirs in Miyun and Huairou Districts were also utilized for flood retention, lowering the water levels of the Chaobai River and facilitating eastward discharge from the Beiyun River; precise scheduling at the Beiguan and Tumenlou Control Complexes directed water into the Chaobai River through the Yunchao and Qinglongwan flood discharge channels, ensuring the safety of the Beiyun River while avoiding the activation of the Dahuangpuwa flood detention area in Tianjin.

Apart from the scientific utilization of the flood control system, levee inspections were also intensified, with 220,000 personnel engaged in levee patrolling and risk assessment and 131 hazards rectified. Through collective efforts, people's lives were protected to the maximum extent, and the safety of critical targets was ensured, with no reservoir failures, no breaches in important levees or embankments of flood detention areas, and zero-casualty evacuation of nearly one million people from flood detention areas. These are the significant achievements in the flood prevention and control battle against the " 23·7 " extreme floods in the Haihe Basin.

案例 2 松花江洪水险情处置

松花江编号洪水期间，拉林河流域磨盘山、龙凤山水库一度超设计洪水位运行；乌苏里江干流全线长历时超保证；拉林河堤防出现多处漫堤溃口，蚂蚁河堤防出现 4 处漫溢、1 处决口。

松辽委针对磨盘山水库、龙凤山水库超设计洪水位运行险情，第一时间增派工作组、专家组赴现场，协助地方开展水文应急监测、应急保坝措施制定和洪水防御等工作，为有效处置险情提供有力技术支撑。

黑龙江省针对磨盘山、龙凤山水库超设计水位运行险情，果断落实超标准洪水应对措施，加强风险研判，保障大流量下泄时河道安全、人员安全，确保水库不垮坝；磨盘山水库溢洪道引水渠前端右侧山体发生泥石流滑坡，大量山石、树木和泥沙涌入水库，利用挖掘机等机械设备及时挖掘清理，确保了溢洪道安全运行；紧急转移龙凤山水库辖区影响人员 111 人，其中集中安置 60 人、分散安置 51 人，无人员伤亡。针对乌苏里江长历时超保证泄水，鸡西市 24 小时调度乌苏里江沿岸各属地开展巡堤查险；虎林市组建 44 支、5600 人应急抢险队伍，抢筑子堤 6.74 千米，处理管涌 6 处，压渗堤防 3.14 千米，加固堤防 11.5 千米，抢修堤防道路 16.3 千米等，累计消除隐患点 37 处，确保"三次洪峰"顺利过境；双鸭山市派出专家督导组到饶河县督导乌苏里江度汛工作，饶河县累计出动 5442 余人次巡查堤防 2240 千米，组建 48 支 3503 人抗洪抢险队伍，加固堤防 39.27 千米，垒筑子堤 29.58 千米；佳木斯市派出 2 个工作组 11 人检查指导防汛工作，派出 77 组 361 人开展不间断巡堤查险，抚远市累计出动 1.8 万余人次巡查堤防 1.7 万千米，落实应急救援队伍 121 支 2358 人，提前转移安置人员 198 人，垒筑子堤 4.88 千米，加固加高 3.16 千米。

吉林省在暴雨笼罩区内小型水库预置挖掘机、彩条布等物料，按照拉林河超标准洪水防御预案，及时组织转移受威胁群众 6.4 万余人，做到应转尽转；督促指导强降水区内漫坝风险较高的水库提前开挖非常溢洪道，发生漫坝的舒兰市谢家店和榆树市水泉水库由于提前对坝顶进行了混凝土硬化，并及时开挖非常溢洪道加大泄流，守住了水库不垮坝的底线；前线工作组与后方指挥部协调联动，连夜指导做好拉林河扶余市蔡家沟镇堤防背水坡回淘塌方处置、花园口段河道内阻水桥爆破清障工作，险情得以有效控制。

Case 2 Hazards handling of the Songhua River floods

During the numbered floods of the Songhua River, Mopanshan and Longfengshan reservoirs in the Lalin River basin temporarily operated above their designed flood handling capacity; the mainstream Wusuli River experienced prolonged and full-line flooding beyond its safe water levels; multiple cases of overflow-induced embankment breach occurred along the Lalin River; four cases of overflow and one breach were recorded at the embankments of the Mayi River.

In response to the critical situation of the Mopanshan and Longfengshan Reservoirs operating above their designed flood levels, the Songliao Commission promptly dispatched working and expert groups to the sites. These teams assisted local authorities in emergency hydrological monitoring, the development of emergency dam protection measures, and flood control efforts, providing strong technical support for effective hazard rectification.

Faced with the hazards at the Mopanshan and Longfengshan reservoirs, Heilongjiang Province decisively took measures to address the excess flood. Risk assessments were enhanced to ensure the safety of the river channels, reservoirs, and people during large-scale flood discharge. A landslide occurred on the right side in front of the spillway channel of the Mopanshan Reservoir, causing a massive influx of rocks, trees, and sediment into the reservoir. Excavation and cleanup operations using mechanical equipment like excavators were promptly conducted to ensure the safe operation of the spillway. A total of 111 people within the impact area of the Longfengshan Reservoir were evacuated, with 60 of them relocated to one place and 51 spread in different locations. No casualty was caused.

Regarding the prolonged, beyond-safe-level floods in the Wusuli River, Jixi City conducted round-the-clock patrols of the embankments in coordination with local governments. Hulin City organized 44 emergency response and rescue teams of 5,600 people, constructed 6.74 km of secondary embankments, dealing with six instances of piping, reinforcing 3.14 km of embankments against seepage, strengthening 11.5 km of embankments, and repairing 16.3 km of levee roads; a total of 37 hazards were eliminated, ensuring the smooth passage of three flood peaks. Shuangyashan City dispatched expert teams to Raohe County to oversee the flood control and prevention operations along the Wusuli River; Raohe County mobilized over 5,442 person-times to inspect 2,240 km of embankments and formed 48 teams comprising 3,503 personnel for flood control and emergency responses, reinforcing 39.27 km of embankments and constructed 29.58 km of secondary embankments. Jiamusi City sent two working groups totaling 11 people to check and guide flood prevention efforts, and dispatched 77 teams of 361 people to carry out non-stop embankment patrols, while Fuyuan City mobilized over 18,000 person-times to patrol the embankments, traveling 17,000 km in total, organized 121 emergency rescue teams totaling 2,358 people, preemptively relocated 198 people, constructed 4.88 km of secondary embankments, and reinforced and raised the height of another 3.16 km.

In Jilin Province, excavators and tarpaulins were pre-positioned at small reservoirs within the area affected by torrential rains; following the Lalin River's contingency plans for extreme flooding, authorities promptly organized the relocation of over 64,000 people, ensuring no affected person was left behind. For reservoirs prone to overtopping due to heavy rainfall, preemptive measures were urged to be taken, such as the excavation of emergency spillways. The Xiejiadian Reservoir of Shulan City and the Shuiquan Reservoir of Yushu City managed to maintain structural integrity through reinforcement of the dam crests with concrete and timely construction of emergency spillways to increase outflow. Thanks to the coordinated efforts of the flood control headquarter and on-site work groups, hazards such as the back-slope scouring and collapse at the embankments of Caijiagou Town in Fuyu City and the water-blocking bridge in need of blasting clearance in the Huayuankou section were effectively dealt with and kept under control.

案例 3 四川芦山县依托测雨雷达开展山洪灾害预报预警

8月6日，四川雅安芦山县出现暴雨天气过程，最大1小时降水量芦山县宝盛站88.5毫米。受强降水影响，芦阳街道西川河大岩腔段、白虎鹰沟暴发山洪。芦山县水利局利用测雨雷达，精准靶向预警沟道上游山洪灾害，提前转移白虎鹰沟等沟道涉水游玩群众1000余人。

8月6日8时许，芦山县气象台发布暴雨蓝色预警，最大小时雨强可达40~70毫米。11时50分，测雨雷达探测到周边宝兴县、邛崃市境内有强对流云团生成，县水利局立即联合县气象局发布山洪灾害气象风险预警，并要求飞仙、芦阳、思延、双石4个镇（街道）强化雨前排查和人员撤离工作。15时许，测雨雷达探测到芦阳街道白虎鹰沟上游无人区出现维持35分钟极强回波，反演小时雨强可达100毫米左右，芦山县水利局立即向芦阳街道办发出山洪预警。

芦阳街道办接到预警后，立即通知村民并组织劝离在西川河大岩腔段、白虎鹰沟、玉溪河程家坝至南门桥段涉水游玩群众千余人。白虎鹰沟100余名游客撤离后不到10分钟，沟道河水暴涨，山洪冲入民房。由于第一时间掌握降水趋势，有效实现靶向预警，街道、社区高效联动，组织村组干部和志愿者快速劝离游客、转移沟边低洼区域群众，使千余群众成功避险。

Case 3 Weather-radar-assisted flash flood disaster forecast and early warning in Lushan County, Sichuan Province

On August 6, rainstorms hit Lushan County, Ya'an City, Sichuan Province, with the maximum one-hour rainfall reaching 88.5 mm at the Baosheng Station. Due to the downpour, flash floods erupted at the Dayanqiang section of the Xichuan River and Baihuying Gully (within the jurisdiction of Luyang neighborhood). With the help of weather radar, the water resources bureau of the county managed to make precise forecast and issue early warnings for the flash flood disasters upstream, and preemptively evacuated over 1,000 tourists from the gullies.

At around 8:00 on August 6, the county meteorological station issued a blue alert for rainstorm, predicting peak hourly rainfall at between 40 and 70 mm. At 11:50, the weather radar detected the formation of strong convective cloud clusters in the neighboring Baoxing County and Qionglai City. The local water resources bureau, in collaboration with the Meteorological Bureau, immediately issued a warning for flash floods, requiring the towns and sub-districts of Feixian, Luyang, Siyan, and Shuangshi to intensify precautionary inspections and evacuations. At around 15:00, the weather radar detected a very strong echo lasting 35 minutes in the uninhabited upper reaches of Baihuying Gully, based on which the hourly rainfall was estimated at around 100 mm. The water resources bureau immediately issued a flash flood alert to the Luyang Neighborhood Office.

Upon receiving the warning, the Luyang Neighborhood Office promptly notified local villagers and organized the evacuation of over a thousand tourists from the Dayanqiang section of the Xichuan River, Baihuying Gully, and the section between Chengjiaba and Nanmenqiao of the Yuxi River. Less than 10 minutes after the evacuation of over 100 tourists from the Baihuying Gully, the river water surged and flash floods gushed into residential buildings. Thanks to the timely grasp of precipitation trends and targeted early warnings, the sub-district and local communities, working efficiently in coordination, swiftly persuaded tourists to leave and relocated residents from low-lying areas adjacent to the gullies, ensuring the safety of over one thousand people.

3.1 旱情

2023 年，全国有 19 省（自治区、直辖市）出现不同程度干旱，西南地区发生春旱、北方局地发生夏旱、西北东部北部发生伏秋旱。全国旱情有 3 个主要特点。

（1）全国旱情总体偏轻。全国因旱农作物受灾面积、绝收面积、粮食损失、饮水困难人口和大牲畜数量，较近 10 年平均值偏少 5～8 成，其中绝收面积、粮食损失和饮水困难人数为近 10 年最少，受灾面积和饮水困难大牲畜头数为近 10 年第 2 少。

3.1 Droughts

In 2023, 19 provinces/autonomous regions/municipalities nationwide experienced droughts of varying degrees. Southwest and parts of North China suffered from a spring drought and a summer drought, respectively, while the east and the north of Northwest China went through droughts stretching from mid-summer through autumn. In general, drought in 2023 took on the following three characteristics.

(1) The drought losses were moderate in general. The cropland area affected or failed by drought, the grain yield loss attributed to drought, the population and the number of bigger-sized livestock having difficulties accessing drinking water were 50%-80% less than the average in the past decade; in particular, the failed cropland area, the grain yield loss and the population having difficulties accessing drinking water reached the lowest in the recent decade, the affected cropland area and the number of bigger-sized livestock having difficulties accessing drinking water were the second lowest in the recent decade.

（2）西南西北旱情重。云南省发生1961年以来最严重干旱，1—4月干旱日数达106天，旱情覆盖全部16市（州）106县（区）[占全省县（区）数的82%]，一度有54万人因旱饮水困难。甘肃省河西走廊发生重现期为60年的旱情且正值秋粮作物生长关键期，威胁粮食生产，甘肃、宁夏2省（自治区）多个城镇供水紧张。内蒙古中西部地区旱情持续近5个月，牧区草场受旱面积大、大牲畜因旱饮水困难问题突出。

（3）旱情多发生在易旱地区。西南地区春旱、北方局地夏旱和西北地区伏秋旱均发生在我国旱灾多发易发地区，呈集中连片特点，与我国干旱规律较为吻合。

3.2 主要干旱过程

3.2.1 西南地区春旱

受2022年冬季降水偏少影响，西南地区云南、贵州、四川3省库塘蓄水持续不足，年初蓄水量总体较常年同期偏少1成。2023年1—4月，西南南部等地降水量较常年同期偏少5～9成，其中云南省平均降水量仅38毫米，为近60年来同期最少；元江、怒江、赤水河来水量较常年同期偏少4～5成；局地中小型水库蓄水严重不足，云南、贵州2省分别有583座、423座水库低于死水位运行。受降水、来水偏少和蓄水不足影响，云南、贵州、四川3省旱情露头并持续发展，部分山丘区和以山塘、泉水等为水源的分散供水片区群众饮水困难问题较为严重，部分高岗地农作物灌溉受到影响。5月末旱情高峰期，3省耕地受旱面积393千公顷，有55万人、20万头大牲畜因旱饮水困难。6月，西南地区陆续出现较强降水过程，加上抗旱措施有力有效，贵州省、四川省旱情解除，云南省旱情逐步缓解。

(2) Droughts were severe in Southeast and Northwest China. Yunnan Province was stricken by the worst drought since 1961, with a long dry period of 106 days from January to April, affecting 16 cities/autonomous prefectures and 106 counties/municipalities (82% of the total counties/municipalities in this province), 540,000 people experienced temporary drinking water shortage. At the Hexi Corridor in Gansu Province, a drought with a 60-year return period occurred during the critical growth period of autumn crops, threatening food production and causing water supply stress in many cities/towns in Gansu Province and Ningxia Autonomous Region. The drought in the middle and western Nei Mongol lasted for nearly 5 months, affecting extensive pastoral grasslands and causing severe drinking water disruptions to the bigger-sized livestock.

(3) Drought occurred mostly in drought-prone areas. Spring drought in Southwest China, summer drought in parts of North China and drought during mid-summer and autumn in Northwest China all occurred in drought-prone and frequently affected areas of the country, showing a pattern of concentration and centralization that is consistent with the drought pattern in China.

3.2 Major Drought Processes

3.2.1 Spring drought in Southwest China

Due to less-than-normal precipitation during winter in 2022, there was a sustained lack of reservoir storage in Yunnan, Guizhou and Sichuan Provinces of Southwest China, with storage at the start of the year 10% less than normal. During January to April in 2023, the southern part of Southwest China received 50%-90% less rainfall than normal; in particular the average rainfall in Yunnan Province was only 38 mm, the lowest in the past six decades. The inflow to Yuanjiang, Nujiang, and Chishui rivers was 40%-50% leaner than normal. Middle and small reservoirs in some regions were not severely under-replenished, with 583 and 423 reservoirs in Yunnan and Guizhou provinces, respectively, not even reaching the dead storage level. Due to the combined shortage of rainfall, river inflow and reservoir storage, drought emerged and developed quickly in the three provinces; drinking water difficulties were pronounced in certain mountainous areas and in regions that rely on decentralized water supply sources such as ponds and springs; some of the upland crops experienced irrigation disruption. Late May was the driest period in the three Provinces: 393,000 ha of cropland were affected, and 550,000 people and 200,000 bigger-sized livestock had difficulties accessing drinking water. In June, as strong rainfall events occurred in Southwest China and drought relief measures began to take effect, drought was subdued in Guizhou and Sichuan provinces and gradually mitigated in Yunnan Province.

3.2.2 北方局地夏旱

5月中旬至6月，华北北部、东北西南部、西北大部、黄淮中北部等地降水量较常年同期偏少4～8成，其中河北省6月平均降水量仅29毫米，为1956年以来同期最少；河北、内蒙古、山东等地气温异常偏高，最高气温较常年同期偏高1～3℃；海河流域拒马河、辽河干流来水较常年同期偏少6～9成。高温少雨和来水不足导致河北、内蒙古、山东、辽宁4省（自治区）旱情快速发展，对农业生产和畜牧养殖业造成影响。6月末至7月初旱情高峰期，4省（自治区）耕地受旱面积3784千公顷，牧区草场受旱面积32667千公顷，有7万人、143万头大牲畜因旱饮水困难。河北承德、张家口等地旱情较重，有12县（市、区）164乡（镇）出现旱情，玉米、大豆、马铃薯等作物生长受较大影响。7月，华北大部、东北西南部陆续出现降水过程，河北、山东、辽宁3省旱情解除，内蒙古东部旱情逐步缓解、西部部分地区旱情持续至9月。

3.2.3 西北地区伏秋旱

6—8月，西北东部北部等地降水量较常年同期偏少3～7成，黄河上游干流、黑河、清水河来水量较常年同期偏少2～3成，西北东部北部水库蓄水总量较常年同期偏少1～2成，甘肃省内陆河流域部分水库趋近干涸。8月上旬，内蒙古、甘肃、青海、宁夏4省（自治区）相邻集中连片地区出现严重旱情，耕地受旱面积1214千公顷，牧区草场受旱面积24667千公顷，甘肃张家川、山丹、古浪和宁夏隆德、盐池、沙坡头等县（区）出现供水紧张，偏远牧区有3万人、74万头大牲畜因旱饮水困难。9月，旱区出现多次降水过程，旱情逐步解除。

3.2.2 Summer drought in parts of North China

From mid May to June, the precipitation in the north of Northeast China, the southwest of Northeast China, most parts of Northwest China, and the central and north of Huanghuai (between the Yellow River and the Huaihe River) Plain was 40%-80% less than normal. In particular, the average rainfall in June in Hebei Province was only 29 mm, the lowest over the same period since 1956; and temperatures in Hebei, Nei Mongol and Shandong provinces/autonomous region were unusually high, with the annual extreme 1-3℃ higher than normal; river inflow to the mainstreams of Juma River and Liaohe River in the Haihe River basin was 60%-90% less than normal. Due to high heat, low precipitation and insufficient river inflow, the drought in Hebei, Nei Mongol, Shandong and Liaoning provinces/autonomous region developed rapidly, affecting agricultural production and stockbreeding. At the peak of drought from late June to early July, 3,784,000 ha and 32,667,000 ha of cropland and pasture were affected, respectively, and 70,000 people and 1.43 million bigger-sized livestock experienced difficulties accessing drinking water. Chengde and Zhangjiakou in Hebei Province took the hardest hit: 164 townships/towns in 12 counties/cities/districts were stricken with drought and the growth of maze, soy beans and potatoes was damaged. In July, most parts of North China and the southwest of Northeast China began to receive rainfall; the drought in Hebei, Shandong, and Liaoning was eliminated, the drought in the east of Nei Mongol was gradually mitigated, and the west of Nei Mongol weathered through localized drought until September.

3.2.3 Mid-Summer and autumn drought in Northwest China

From June to August, the east and north of Northeast China was 30%-70% drier than normal; the river inflow to mainstream of the upper Yellow River, the Heihe River and the Qingshui River was 20%-30% less than normal; the total water storage of reservoirs in the east and north of Northwest China was 10%-20% less than normal; some of the reservoirs in the inland river basins of Gansu Province were nearly dry. In early August, severe droughts appeared in centralized and adjacent areas in the four Provinces/autonomous region of Nei Mongol, Gansu, Qinghai, and Ningxia, affecting 1,214,000 ha of cropland and 24,667,000 ha of pasture. In particular, counties/districts such as Zhangjiachuan, Shandan and Gulang in Gansu and Longde, Yanchi, Shapotou in Ningxia began to experience water supply stress, and 30,000 people and 740,000 bigger-sized livestock in remote pastoral areas experienced drinking water disruptions. In September, the drought was gradually eliminated after several rounds of rainfall.

此外，珠江流域 2022 年 9 月至 2023 年 3 月河口地区出现 11 轮强咸潮，影响时间长达 7 个月，为有监测记录以来最长，珠海平岗泵站平均抽取淡水概率为 2005 年有咸情监测记录以来最低；2023 年 4—10 月，珠江流域江河来水偏少 2～5 成，其中西江偏少近 5 成，为 1950 年有完整实测资料以来同期第 2 少。珠江下游枯水和咸潮期间，澳门、珠海等地供水安全受到威胁。长江干流 6—8 月来水量较常年同期偏少 4 成，7 月下旬中下游干流及洞庭湖、鄱阳湖水位较常年同期偏低 4.57～7.76 米，湖北、湖南、云南、四川等地插花受旱。

3.3 干旱灾情

2023 年，全国 19 省（自治区、直辖市）发生干旱灾害。农作物因旱受灾面积 3803.70 千公顷，比前 10 年平均值下降 54.3%；绝收面积 218.37 千公顷，比前 10 年平均值下降 76.4%；因旱粮食损失 33.88 亿千克，比前 10 年平均值下降 75.0%；有 274.74 万人因旱发生饮水困难，比前 10 年平均值下降 67.9%；278.29 万头大牲畜因旱饮水困难，比前 10 年平均值下降 52.8%。其中农作物因旱绝收面积、因旱粮食损失和因旱饮水困难人数为近 10 年最少。

In addition, 11 rounds of strong saltwater intrusion, lasting seven months from September 2022 to March 2023, occurred at the Pearl River estuary. It was the longest duration since monitoring began. The average freshwater abstraction probability at the Pinggang Pumping Station in Zhuhai City reached the lowest since the first record of saltwater intrusion in 2005. During April to October, river inflow to the Pearl River basin was 20%-50% less than normal; in particular the Xijiang River received 50% less inflow than normal, marking the second lowest since complete observation data was available in 1950. During the period when the Pearl River suffered from low instream flow and saline intrusion downstream, water supplies to Macao SAR and Zhuhai were threatened. Inflow to the mainstream Yangtze River was 40% less than normal during June to August; in late July the water level was 4.57-7.76 m lower than normal in the middle and lower reaches of its mainstream as well as Dongting Lake and Poyang Lake; Hubei, Hunan, Yunnan and Sichuan Provinces experienced sporadic and localized droughts.

3.3 Disasters and Losses

In 2023, 19 Provinces/autonomous regions/municipalities nationwide experienced drought disasters. A total of 3,803,700 ha of cropland were affected, 54.3% less than the preceding decadal average; 218,370 ha suffered crop failure, 76.4% lower than the preceding decadal average; the grain yield loss attributed to drought was 3.388 billion kg, down by 75.0% from the preceding decadal average; a total of 2.7474 million people experienced difficulties accessing drinking water due to drought, 67.9% less than the preceding decadal average; and 2.7829 million bigger-sized livestock experienced difficulties accessing drinking water due to drought, down by 52.8% from the preceding decadal average. It is noted that the affected cropland area suffering crop failure, the grain yield loss and the population experiencing difficulties accessing drinking water due to drought marked their lowest in the past decade.

表 3-1 2023 年全国农作物因旱受灾面积、绝收面积情况（单位：千公顷）
Table 3-1 Cropland area affected and failed by drought in 2023 (in 1,000 ha)

地区 Province	农作物因旱受灾面积 Cropland area affected by drought	农作物因旱绝收面积 Cropland area failed by drought	地区 Province	农作物因旱受灾面积 Cropland area affected by drought	农作物因旱绝收面积 Cropland area failed by drought
全国 Nationwide	3803.70	218.37	河南 Henan		
北京 Beijing			湖北 Hubei	38.73	1.09
天津 Tianjin			湖南 Hunan	151.16	9.45
河北 Hebei	225.07	23.93	广东 Guangdong		
山西 Shanxi	175.41	11.53	广西 Guangxi	107.19	8.24
内蒙古 Nei Mongol	1526.86	51.54	海南 Hainan		
辽宁 Liaoning	256.10	2.75	重庆 Chongqing	12.62	0.93
吉林 Jilin			四川 Sichuan	126.58	7.66
黑龙江 Heilongjiang			贵州 Guizhou	105.97	7.83
上海 Shanghai			云南 Yunnan	503.02	26.54
江苏 Jiangsu			西藏 Xizang		
浙江 Zhejiang	0.65		陕西 Shaanxi	310.15	38.49
安徽 Anhui			甘肃 Gansu	211.04	17.35
福建 Fujian			青海 Qinghai	14.40	0.64
江西 Jiangxi			宁夏 Ningxia	37.94	10.40
山东 Shandong			新疆 Xinjiang	0.81	

注 数据来源于应急管理部，空白表示无灾情。
Note The data come from the Ministry of Emergency Management, and spaces in blank denote no losses or damages.

表 3-2 2023 年全国因旱饮水困难情况

Table 3-2 Difficulties accessing drinking water attributed to drought nationwide in 2023

地区 Province	因旱饮水困难人口/万人 Population having difficulties accessing drinking water /10,000 persons	因旱饮水困难大牲畜/万头 Number of bigger-sized livestock having difficulties accessing drinking water/10,000 heads	地区 Province	因旱饮水困难人口/万人 Population having difficulties accessing drinking water /10,000 persons	因旱饮水困难大牲畜/万头 Number of bigger-sized livestock having difficulties accessing drinking water/10,000 heads
全国 Nationwide	274.74	278.29	河南 Henan		
北京 Beijing	3.38		湖北 Hubei	4.08	1.64
天津 Tianjin			湖南 Hunan	1.87	0.34
河北 Hebei	5.23	1.92	广东 Guangdong		
山西 Shanxi	3.90	1.42	广西 Guangxi	14.90	2.67
内蒙古 Nei Mongol	4.96	99.40	海南 Hainan	0.30	
辽宁 Liaoning	0.19	0.02	重庆 Chongqing	11.69	8.72
吉林 Jilin			四川 Sichuan	37.24	9.57
黑龙江 Heilongjiang			贵州 Guizhou	14.55	1.30
上海 Shanghai			云南 Yunnan	155.54	50.51
江苏 Jiangsu			西藏 Xizang		
浙江 Zhejiang			陕西 Shaanxi	9.96	1.05
安徽 Anhui			甘肃 Gansu		35.02
福建 Fujian			青海 Qinghai	6.93	24.79
江西 Jiangxi			宁夏 Ningxia		39.52
山东 Shandong			新疆 Xinjiang	0.02	0.40

图 3-1 2023 年全国干旱灾害分布图
Figure 3-1 Overview of drought disasters nationwide in 2023

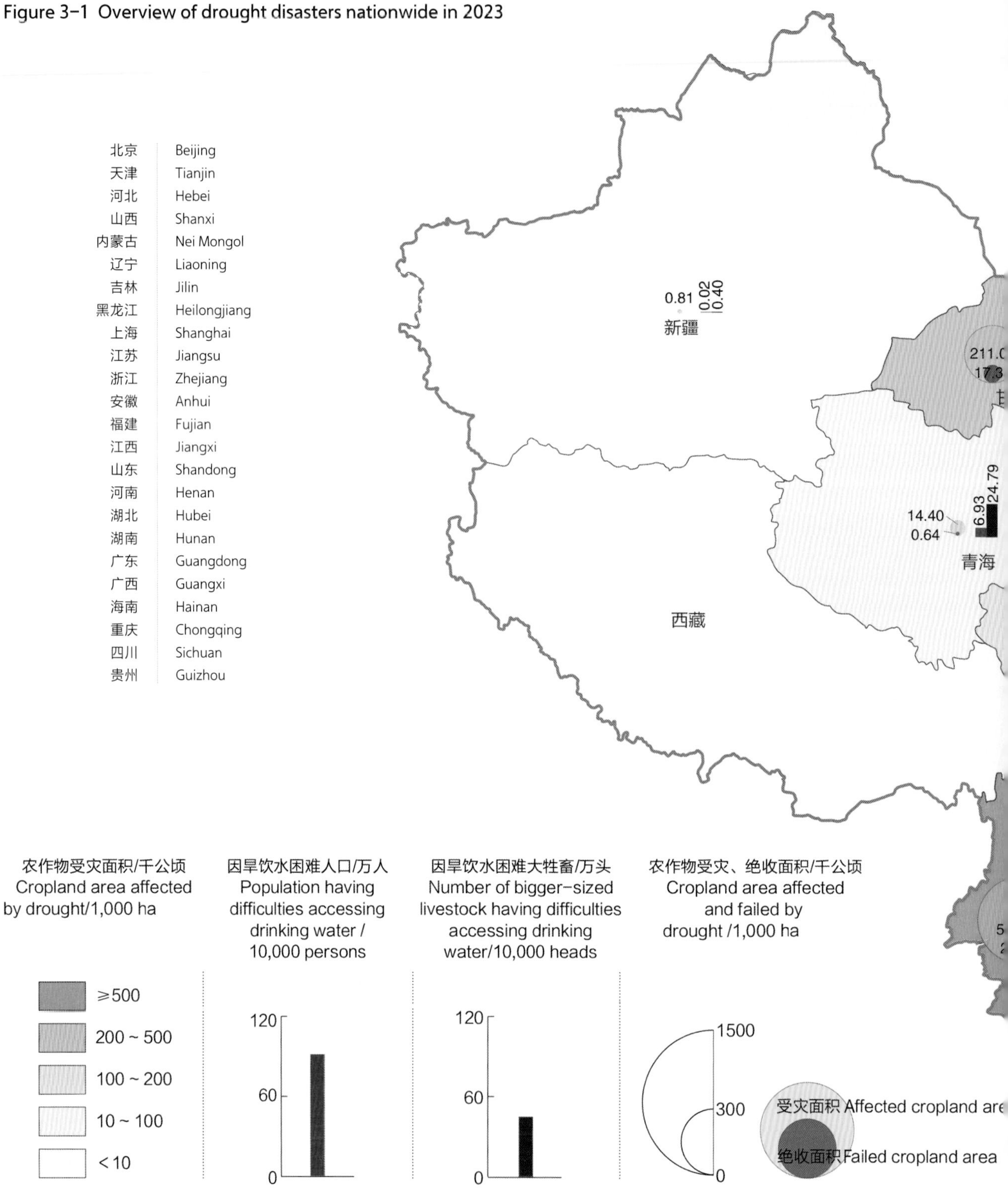
北京 Beijing
天津 Tianjin
河北 Hebei
山西 Shanxi
内蒙古 Nei Mongol
辽宁 Liaoning
吉林 Jilin
黑龙江 Heilongjiang
上海 Shanghai
江苏 Jiangsu
浙江 Zhejiang
安徽 Anhui
福建 Fujian
江西 Jiangxi
山东 Shandong
河南 Henan
湖北 Hubei
湖南 Hunan
广东 Guangdong
广西 Guangxi
海南 Hainan
重庆 Chongqing
四川 Sichuan
贵州 Guizhou
0.81
0.02
0.40
新疆
西藏
14.40
0.64
6.93
24.79
青海
农作物受灾面积/千公顷 Cropland area affected by drought/1,000 ha
≥500
200 ~ 500
100 ~ 200
10 ~ 100
<10
因旱饮水困难人口/万人 Population having difficulties accessing drinking water / 10,000 persons
120
60
0
因旱饮水困难大牲畜/万头 Number of bigger-sized livestock having difficulties accessing drinking water/10,000 heads
120
60
0
农作物受灾、绝收面积/千公顷 Cropland area affected and failed by drought /1,000 ha
1500
300
0
受灾面积 Affected cropland are
绝收面积 Failed cropland area

注 香港特别行政区、澳门特别行政区、台湾省资料暂缺。
Note The data of Hong Kong SAR, Macao SAR and Taiwan are currently unavailable.

3 干旱灾害防御

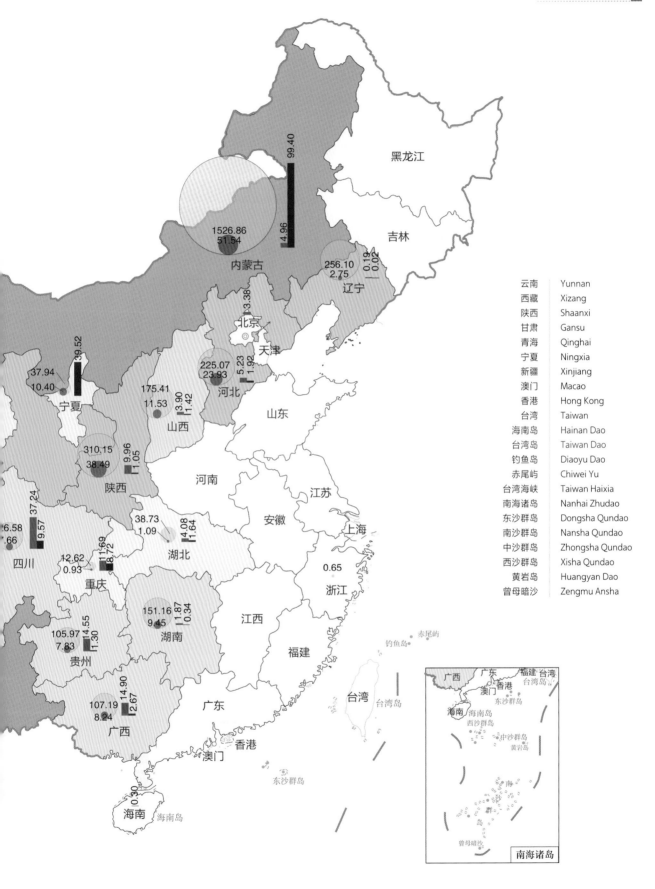

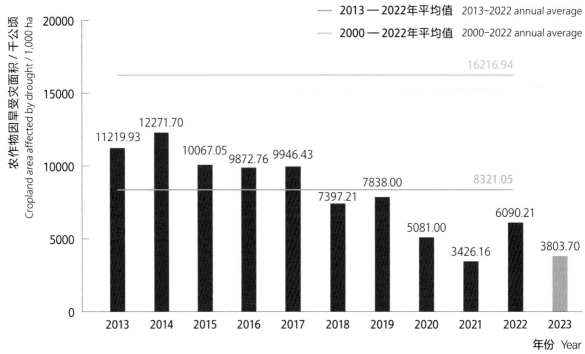

注 2019—2023 年数据来源于应急管理部。
Note The data during 2019-2023 come from the Ministry of Emergency Management.

图 3-2 2013—2023 年全国农作物因旱受灾面积
Figure 3-2 Cropland area affected by drought 2013-2023

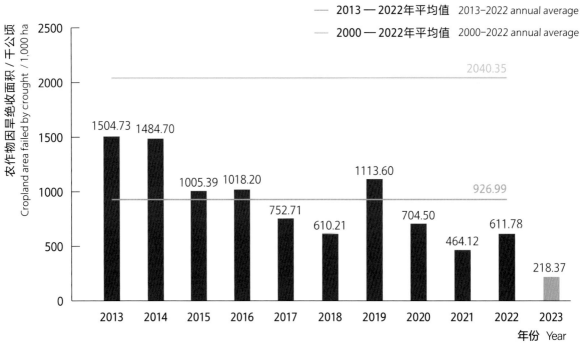

注 2019—2023 年数据来源于应急管理部。
Note The data during 2019-2023 come from the Ministry of Emergency Management.

图 3-3 2013—2023 年全国农作物因旱绝收面积
Figure 3-3 Cropland area failed by drought 2013-2023

3 干旱灾害防御

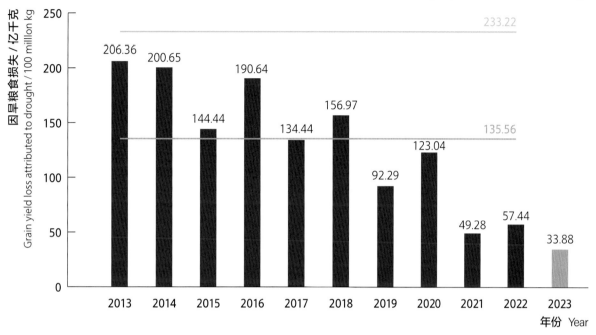

图 3-4 2013—2023 年全国因旱粮食损失
Figure 3-4　Grain yield loss attributed to drought 2013-2023

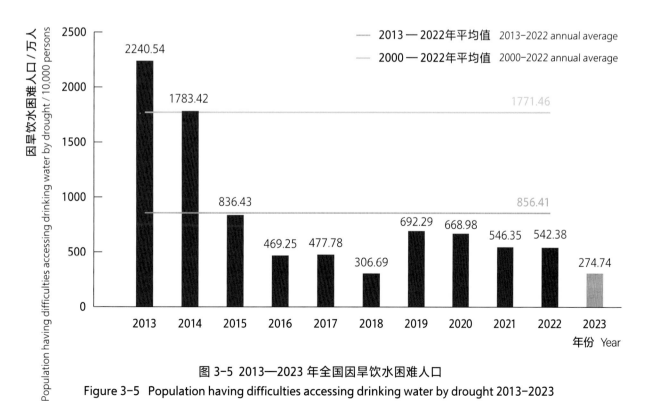

图 3-5 2013—2023 年全国因旱饮水困难人口
Figure 3-5　Population having difficulties accessing drinking water by drought 2013-2023

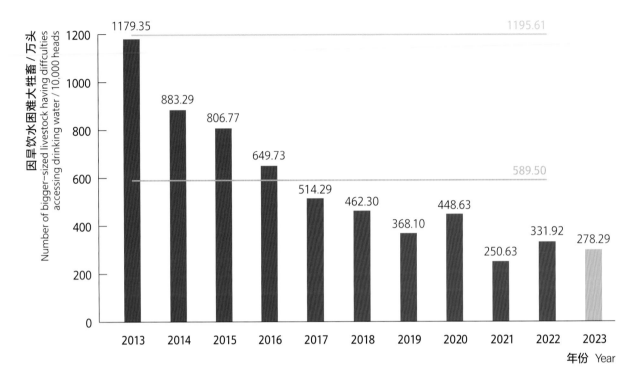

图 3-6 2013—2023 年全国因旱饮水困难大牲畜

Figure 3-6 Number of bigger-sized livestock having difficulties accessing drinking water by drought 2013-2023

3.4 防御工作

水利部坚持人民至上、生命至上，统筹发展和安全，组织指导旱区精准范围、精准对象、精准时段、精准措施，有序做好各项抗旱工作，确保城乡群众饮水安全，确保农作物时令灌溉用水需求。

3.4 Prevention and Control

MWR kept protecting the people and their lives at the first place, coordinated the needs of development and safety, and commanded and guided the drought-stricken areas to fight against and relieve droughts with a precise judgement on scope, targets, timing, and actions to be taken. Urban and rural drinking water safety as well as crop irrigation needs at the time were protected.

3.4.1 工作部署

李国英部长多次会商部署抗旱工作，8月15日主持抗旱专题会商会，视频连线黄委，研究部署西北地区抗旱保供水工作。水利部编制《水利部贯彻落实习近平总书记关于未雨绸缪做好农业防灾减灾重要指示实施方案》，细化实化任务分工，落实各项工作措施；加大对受旱地区指导支持力度，先后发出6个通知部署抗旱工作，派出8个工作组赴云南、贵州、河北、内蒙古、甘肃、宁夏、青海等旱区一线，现场察看旱情。水利部及黄委针对内蒙古、甘肃、青海、宁夏4省（自治区）旱情分别启动干旱防御Ⅳ级应急响应，组织指导受旱地区全力做好抗旱工作。河北、内蒙古、山东、广西、贵州、云南、甘肃、宁夏等省（自治区）及时启动干旱防御Ⅳ级应急响应（其中云南省应急响应持续85天），旱区党委、政府主要负责同志对抗旱作出安排部署，并赴一线指导。

3.4.1 **Arrangements**

Minister Li Guoying guided the drought prevention and control through multiple consultations. On August 15, the Minister convened the online meeting with the Yellow River Commission where the works related to drought relief and water supply protection were discussed and deployed. MWR has compiled the *Implementation Plan on Executing the Key Central Instructions on Proactive Disaster Prevention and Reduction in Agriculture*, refined and specified the responsibilities and implemented various preventive measures. With greater support and guidance to the drought-affected areas, the Ministry issued six notices on the deployment of drought mitigation, and sent eight working groups to the front line of the drought-stricken regions including Yunnan, Guizhou, Hebei, Nei Mongol, Gansu, Ningxia, and Qinghai. Level Ⅳ emergency responses against drought was launched by MWR and the Yellow River Commission in Nei Mongol, Gansu, Qinghai and Ningxia, and all-out efforts were made to guide the drought relief work. Hebei, Nei Mongol, Shandong, Guangxi, Guizhou, Yunnan, Gansu and Ningxia launched level Ⅳ emergency responses, among which the emergency responses lasted for 85 days in Yunnan. Local government officials in charge were attentive to the arrangements made on drought relief and visited the affected areas for instruction.

3.4.2 "四预"措施

水利部强化抗旱"四预"措施，牢牢掌握抗旱工作的主动权。密切监视全国雨情、水情、墒情、旱情，滚动预测预报，加强会商研判，及时发布预警，全年共发布51次干旱预警；组织各流域管理机构编制应急水量调度预案，分析流域供水形势并提出供水保障措施。

各地落实抗旱"四预"措施，科学研判旱情态势，针对性制定各项抗旱保供水措施。内蒙古自治区加密旱情监测分析预报，及时准确掌握供水紧张和饮水困难情况，为抗旱决策提供支撑。江西省坚持"三个10天"旱情预警机制，预报预测未来10天、20天、30天的旱情，精准测算农作物需水情况，精细谋划保供水措施。重庆市建立"村组排查、乡镇巡查、县级督查、市级指导"的4级联动监测机制，对供水保障情况开展日监测、日巡查、日调度，发布抗旱保供工作提示，提醒相关地区做好抗旱工作。云南省发布干旱预警146条、河道枯水位预警4次，组织开展省市县乡村5级供需水平衡分析，在算清水账基础上精准制定保供水方案。甘肃省重点针对农作物灌溉缺水、因旱人饮困难和城市干旱缺水等问题突出的区域，编制《甘肃省水利厅抗旱方案（2023年9月—2024年3月）》，落实相关工作措施。河北、吉林、安徽、重庆、宁夏等省（自治区、直辖市）编制完成省级行政区域应急水量调度预案，广东省编制完成东江、韩江等江河应急水量调度预案，四川省编制完成沱江、涪江等江河应急水量调度预案，为流域、区域开展应急水量调度工作提供重要依据。

3.4.2 The "four preemptive pillars"

MWR has shored up the "four preemptive pillars" (forecasting, early warning, exercising and contingency planning) to prevent and fight drought; closely monitored rainfall, water regime, soil moisture and drought conditions; conducted non-stop forecasting and prediction; enhanced analysis and judgement through consultations; issued early warnings in a prompt manner (drought alerts for 51 times in 2023); coordinated the river basin commissions to draft emergency water dispatch plan to assess water supply situation and propose measures accordingly.

The "four preemptive pillars" were efficiently implemented at local levels by scientifically analyzing the trend of drought conditions and taking targeted actions on drought mitigation and water supply. Nei Mongol intensified the monitoring, analysis and forecast of droughts, in a bid to obtain well-timed and accurate information regarding water supply stress and drinking water difficulties, thus supporting drought-related decision making. Jiangxi Province continued with a drought early-warning mechanism called "three 10 days", namely, continuously forecast the trend of drought for the next 10, 20 and 30 days; calculated the exact water needs of crops and planned the water supply measures in great detail. Chongqing Municipality established a 4-level monitoring mechanism of "village-level investigation, town-level inspection, county-level supervision and city-level guidance"; undertook a day-by-day monitoring, inspection and dispatch on water supply; and called for full implementation of drought mitigation and water supply measures. Yunnan Province issued 146 drought early warning messages and 4 times of low water level warning; conducted analysis on water supply-demand balance at Province, city, county, town and village levels, and made informed water supply plan based on water accounting. A Drought Mitigation Plan (September 2023-March 2024) was drafted by Gansu Provincial Water Department, focusing on areas threatened by irrigation water scarcity, drinking water difficulties and urban water scarcity. Provinces/autonomous regions/municipalities including Hebei, Jilin, Anhui, Chongqing, and Ningxia formulated emergency water dispatch plans at provincial level. Guangdong Province compiled the emergency water dispatch plans of Dongjiang and Hanjiang rivers; Sichuan Province compiled the emergency water dispatch plans of Tuojiang and Fujiang rivers, providing an important basis for emergent water dispatch within the basin and the whole region.

3.4.3 抗旱调度

水利部组织指导相关流域管理机构强化抗旱水源统一调度，为抗旱提供稳定水源保障。长江委调度以三峡水库为核心的长江上游水库群，1月至4月中旬向下游补水260亿立方米，汛前在三峡及金沙江梯级等水库留存了30多亿立方米水量，保障沿江城市正常取水和沿线灌区引水；汛末统筹防汛和抗旱需要，三峡、丹江口水库完成满蓄目标，纳入流域联合调度水库群蓄水量创历史新高。黄委调度龙羊峡、刘家峡等骨干水库，在农作物生长关键期维持黄河干流兰州断面流量1000～1300立方米每秒；做好黑河等内陆河水量调配，保障沿河城乡供水和灌区引水等需求。淮委调度淮河流域大型水库后汛期拦蓄雨洪资源近40亿立方米，指导江苏省做好洪泽湖蓄水调度，为秋冬季储备抗旱水源。海委调度引滦工程、岳城水库等水利工程，累计供水14亿立方米，保障天津、河北、河南等省（直辖市）灌溉用水。珠江委2022年冬至2023年春实施8次珠江枯水期压咸补淡应急水量调度，有效压制咸潮，将西北江三角洲磨刀门水道咸界最大下移约41千米，累计向澳门、珠海供水1.54亿立方米，解除咸潮上溯对澳门、珠海等地供水威胁。松辽委加大嫩江流域尼尔基水库春灌期间出库流量，保障了沿江灌区35.33千公顷水田春灌用水；调度察尔森水库向吉林白城市进行生态补水，有效缓解白城市旱情、改善生态环境。太湖局组织实施望虞河、新孟河引江济太应急调水，全年引长江水19.71亿立方米，入太湖6.20亿立方米，保障太湖水源地供水安全。

3.4.3 Water dispatch for drought relief

Led by MWR, relevant river basin commissions reinforced the unified dispatch of water sources, aiming to guarantee reliable water sources to stave off droughts. Upstream reservoir group in the Yangtze River with the Three Gorges as its core were dispatched by the Changjiang Commission and released 26 billion m^3 of water downstream from January to mid April. Before the flood season arrived, over 3 billion m^3 of water was impounded at the Three Gorges and the cascade reservoirs on the Jinsha River, so as to ensure regular water withdrawal by the riparian cities and irrigation districts; by the end of the flood season, to coordinate the works of flood prevention and drought mitigation, the Three Gorges and the Danjiangkou reservoirs were impounded to the full and the total storage by the reservoir group included into the joint dispatch action reached record high. The Yellow River Commission dispatched the backbone reservoirs like Longyangxia and Liujiaxia to maintain a flow rate of 1,000-1,300 m^3/s at the Lanzhou section, mainstream of the Yellow River, during the critical growth period of crops; made reasonable water allocation in inland rivers such as the Heihe River to safeguard urban and rural water supply along the rivers and water diversion in irrigation districts. The Huaihe Commission, by dispatching large reservoirs within the basin, impounded rain and floodwater for nearly 4 billion m^3; and guided Jiangsu provincial government to dispatch water from the Hongze Lake as a reserve of water resources against autumn and winter droughts. The Haihe Commission dispatched water projects including the Luanhe-Tianjin Water Diversion Project and Yuecheng Reservoir, in total diverting 1.4 billion m^3 of water, supporting the irrigation water use in Tianjin, Hebei and Henan. The Pearl River Commission initiated 8 rounds of emergency water dispatch to check saltwater intrusion during the low-water season of the Pearl River from 2022 wintertime to 2023 springtime; the freshwater/saltwater interface retreated by 41 km at the Modaomen waterway in the Xibei River delta; 154 million m^3 of water was dispatched to Macao and Zhuhai to eliminate water supply risks posed by the salt tide. The outflow of the Nierji Reservoir in the Nenjiang River basin was increased by the Songliao Commission during spring irrigation, benefiting 35,330 ha of cropland; the Chaersen Reservoir was dispatched to replenish ecological water requirements to Baicheng City in Jilin Province, effectively alleviating drought and sustaining the local ecosystems. The Taihu Authority conducted emergency water diversion through Yangtze-Taihu Water Diversion Project from the Wangyu and Xinmeng rivers, diverting 1.971 billion m^3 of water from the Yangtze River and replenishing 620 million m^3 of water to the Taihu Lake, backing up the water supply security in the Taihu Lake source water regions.

旱区各地通过水库放水、涵闸引水、泵站提水、渠道输水等综合措施及河湖、湖库、库闸联调等多种手段，全力保障抗旱用水。天津市延长引滦调水时间，保持于桥水库蓄水充足，向州河、蓟运河、潮白新河补充生态及农业用水。内蒙古自治区调度引黄工程应急引水 7.62 亿立方米，有效保障河套灌区 333.33 千公顷小麦、玉米等作物灌溉用水需求。辽宁省调度辽河、浑河、太子河补水 17.81 亿立方米，全力保障 226.67 千公顷灌区小麦等作物灌溉用水。黑龙江省调度大顶子山航电枢纽向下游补水 1.1 亿立方米，保障了沿松花江灌区 5.87 千公顷水田灌溉用水。江西省"龙舟水"期间，抢抓降水有利时机增加蓄水 34.84 亿立方米，台风"杜苏芮"影响期间增蓄 3.48 亿立方米，为应对可能出现的干旱储备了水源。湖北省调度水库、泵站向全省大中型灌区供水 118 亿立方米，其中引江济汉工程向汉江、长湖补水 26 亿立方米，鄂北工程向襄阳、随州、孝感等地供水 1.37 亿立方米，保障了工程沿线群众饮水和农业灌溉用水。湖南省局地干旱期间加大引水提水力度，累计引提水 2.57 亿立方米，有效解决洞庭湖区灌溉缺水问题。四川省累计调水超 50 亿立方米，保障了 1868.67 千公顷水稻栽插用水，完成抗旱浇地 105.33 千公顷，全力稳住粮食生产"基本盘"。云南省提前科学调度水库蓄水保水，至 10 月 31 日全省库塘蓄水 85.86 亿立方米，为秋冬季抗旱保供水奠定了水源基础。甘肃省加大景电、引大、引洮等骨干水利工程调水供水力度，累计引水 10.02 亿立方米，有效缓解河西地区旱情。

Local governments in the drought-affected areas used a portfolio of measures to guarantee water resources during the droughts, including reservoir water, diverted water by culverts and sluices, pumped water, and transported water via channels. Multiple water sources including rivers and lakes, reservoirs, and engineering projects were mobilized. Tianjin prolonged the water diversion period from the Luanhe River to keep sufficient storage in the Yuqiao Reservoir and replenish agricultural and ecological water requirements to the Zhouhe River, Jiyun River, and Chaobai New River. Nei Mongol diverted 762 million m^3 of water from the Yellow River in emergency operations, guaranteeing the irrigation water use in the Hetao Irrigation Area with 333,330 ha of crops such as wheat and maze. Liaoning dispatched water from the Liaohe, the Hunhe and the Taizi rivers, replenishing 1.781 billion m^3 of water to support the irrigation water use of 226,670 ha of cropland. Heilongjiang dispatched water from the Dadingzishan Navigation and Hydropower Project to replenish 110 million m^3 of water downstream, supporting 5,870 ha of cropland in the irrigation areas along the Songhua River. During "Dragon Boat Rainy Season", Jiangxi took the opportunity to increase the water storage by 3.484 billion m^3; when Typhoon Doksuri landed, 348 million m^3 was stored additionally to prepare for drought. Hubei dispatched reservoirs and pumping stations to supply 11.8 billion m^3 of water to mid-and-large irrigation areas within the Province. In particular, relying on the Yangtze-Hanjiang Water Diversion Project, 2.6 billion m^3 of water was replenished to the Hanjiang River and the Changhu Lake; and 137 million m^3 was diverted to Xiangyang, Suizhou and Xiaogan through water resources allocation projects in the north Hubei, securing the drinking water access and irrigation water use in the areas along the project. When parts of Hunan suffered drought, the local government leveled up water extraction to 257 million m^3 in total, effectively solving the irrigation water shortage in the Dongting Lake area. Sichuan diverted over 5 billion m^3 of water in total, providing water for rice planting covering an area of 1,868,670 ha and saving a total of 105,330 ha of affected cropland; food production was saved as a result. Yunnan scientifically dispatched reservoirs in advance for water storage; till October 31 reservoir and pond storage within the province reached 8.586 billion m^3, which served as water sources for drought relief during autumn and winter. Gansu shored up water diversion and supply by major water projects including the Jingtaichuan electrical water lifting for irrigation project, the Dadu-Minjiang water diversion project and the Taohe River water diversion project, etc., in total diverting a 1.002 billion m^3 of water to mitigate the drought in the Hexi area.

3.4.4 抗旱投入

水利部商财政部累计安排中央水利救灾资金 3.454 亿元，支持旱区开展 399 个抗旱应急项目，其中添置提水运水设备和抗旱用油用电等非工程类项目 38 个、引提调水工程 320 个、蓄水工程 41 个，有效提升区域抗旱保供水能力。受旱地区加大抗旱资金投入，推动抗旱项目落地，提升旱区抗旱减灾能力。河北省安排省级财政抗旱补助资金 1000 万元，支持兴建抗旱水源、购买调水供水设施设备，补助调水工程运行投入。山西省安排省级抗旱资金 604 万元，用于修编抗旱方案预案、维修养护抗旱设施、补助抗旱拉运水等。湖北省各地累计投入抗旱资金 3.66 亿元，发动干部群众 12 万人次，启动抗旱泵站 3000 余座、机动设备 2.6 万台套，共挽回粮食损失 9.68 亿元。湖南省各级累计出动抗旱劳力 125 万人，投入抗旱资金 4.9 亿元、抗旱设备 30.4 万台（套），保障 1.87 万农村人口、0.34 万头大牲畜和 100 千公顷农田用水需求。广西壮族自治区及时安排救灾资金 1000 万元，支持 38 个受旱县（市、区）采取兴建抗旱应急水源工程、应急抽水调水、拉水送水等措施保障用水需求。重庆市安排市级抗旱救灾资金 3000 万元，支持建设 81 个抗旱项目，保障 16 万人、6 万头大牲畜和 6.67 千公顷农田灌溉用水需求。云南省投入 12.36 亿元（其中省级抗旱资金 2 亿元），建成 646 项抗旱保供水工程，解决了 187.6 万人饮水困难，提升了 99 千公顷农田灌溉保障能力，挽回粮食损失 2.5 亿千克、经济作物损失约 13.97 亿元。

3.4.4 Financial and in-kind input for drought relief

MWR, in consultation with the Ministry of Finance, allocated 345.4 million RMB of the central water disaster relief fund to support 399 emergency drought relief projects, striving to effectively improve regional capacity in terms of drought mitigation and water supply. Specifically, there were 38 non-engineering projects of purchasing equipment for water extraction and transport and subsidizing the cost of engine oil and electricity consumed for drought prevention, 320 engineering projects for water diversion and pumping, and 41 water storage projects. Drought-affected areas have increased their investment in drought mitigation, fostered the implementation of drought-resistance projects and put efforts in enhancing drought prevention and disaster mitigation capabilities. Hebei Province arranged 10 million RMB from the provincial fiscal coffers as the drought relief fund, endeavoring to support the construction of drought-resistance water sources, purchase facilities and equipment for water diversion and supply, and supplement the operation cost of water diversion projects. Shanxi Province allocated 6.04 million RMB from the provincial fiscal coffers as the drought relief fund, mainly used for revising contingency plans, repairing and maintaining drought-resistance facilities, and subsidizing the cost of water transport, etc. In Hubei Province, a total of 366 million RMB were invested in drought mitigation, 120,000 officials and residents were mobilized cumulatively, over 3,000 pumping stations and 26,000 sets of mechanical equipment were put into operation. As a result, grain yield loss worth of 968 million RMB was saved. In Hunan Province, a total of 1.25 million people participated in drought-resistance actions as labor force, 490 million RMB fund was invested, 304,000 sets of equipment were applied, thus safeguarding the water needs of 18,700 rural population, 3,400 bigger-sized livestock and 100,000 ha cropland. Guangxi autonomous region promptly mobilized 10 million RMB as disaster relief fund to support 38 affected towns/cities/districts to take responsive measures including emergency water sources projects construction, emergency water extraction and diversion, water delivery and transport, in a bid to secure water supply. Chongqing municipality allocated 30 million RMB fund for disaster relief and mitigation, supporting 81 drought-resistance projects that guaranteed the water needs of 160,000 people, 60,000 bigger-sized livestock and 6,670 ha cropland. A total of 1.236 billion RMB was invested in Yunnan Province, of which the provincial drought-resistance fund was 200 million RMB; 646 projects for drought relief and water supply were completed, addressing 1.876 million people's difficulties accessing drinking water, improving the irrigation supplied to 99,000 ha cropland, and recovering 250 million kg of grain yield loss and cash crops loss worth of 1.397 billion RMB.

贵州遵义建设抗旱应急引水工程（2月23日）
An emergency water diversion project for drought relief was built in Zunyi, Guizhou Province (February 23)

湖北宜昌秭归县水窖水池储水，提前做好抗伏旱准备（7月14日）
Water cellars and tanks were filled up in anticipation of a mid-summer drought in Zigui County, Yichang, Hubei Province (July 14)

3.5 防御成效

2023年，旱区各地抗旱累计解决饮水困难人口272.4万人，完成抗旱浇地面积6099.71千公顷，挽回粮食损失64.15亿千克、经济作物损失48.88亿元，确保了城乡居民饮水安全，确保了规模化养殖和大牲畜饮水安全，保障了农作物时令灌溉用水需求，为全年粮食丰产作出贡献。

3.5 Effectiveness of Drought Disaster Prevention

In 2023, local water departments of the drought-stricken areas supplied water to 2.724 million people with difficulties accessing drinking water, completed the emergency irrigation in 6,099,710 ha of cropland. As a result, 6.415 billion kg of grain yield and cash crops worth of 4.888 billion RMB were recovered. Drinking water needs of urban and rural residents as well as large-scale farming and bigger-sized livestock were secured, seasonal irrigation demands by crops were met, thus contributing to the annual grain production.

案例 4 2022—2023 年珠江枯水期压咸补淡应急水量调度

2022—2023年枯水期，西江降水、来水持续偏少，西江干流梧州站天然来水量较常年偏少近5成，为1950年有完整资料以来第2少，西江遭遇特枯水年。截至2023年3月底，珠江河口共出现11轮强咸潮，影响时间为2005年有咸情监测记录以来最长，较常年偏长近1个月；磨刀门水道咸潮最远上溯约53千米，珠海、中山等地主要取水口含氯度超标时间较常年偏早1~2个月，珠海主力取水泵站平岗泵站枯水期平均抽取淡水概率为2005年以来最低。水利部指导珠江委强化流域统一调度，与有关省（自治区）水利、电力、航运、气象等部门密切配合，组织实施8轮压咸补淡应急补水调度，次数之多、历时之长均创历史之最。累计下达64道调度指令，西北江骨干水库群向下游补水76亿立方米，将西江梧州站平均流量提高近5成，磨刀门水道咸界最多下移约41千米，压制了河口咸潮，为下游珠海、中山抢淡蓄水创造了有利条件，累计向澳门、珠海供水1.54亿立方米，其中向澳门供水0.44亿立方米，连续19年确保了粤港澳大湾区城市供水安全，同时保障了西江航运畅通和电网运行安全，改善了河道水生态水环境，实现多方共赢。

Case 4 Emergency freshwater dispatch against saltwater intrusion during low-water season in the Pearl River between 2022-2023

During the low-water season between 2022-2023, rainfall and river inflow of the Xijiang River were constantly low. The natural inflow at the Wuhou Station on Xijiang River was nearly 50% less than normal and the second lowest since the first complete documentation in 1950. It marked an extremely dry year in the Xijiang River. By the end of March 2023, 11 rounds of strong saltwater intrusion occurred at the Pearl River estuary, almost one month longer than normal in terms of the duration, marking the longest duration since recording began in 2005. At the Modaomen waterway in the Xijiang River, the salt tide ran upstream for about 53 km in its longest. At the main water intakes in cities like Zhuhai and Zhongshan, excessive chloride content was detected 1-2 months earlier than normal; Pinggang Pumping Station, one of the major pumping stations in Zhuhai, saw its record-low average freshwater extraction probability during dry season since 2005. Guided by MWR, the Pearl River Commission scaled up the unified coordination of the whole basin, worked closely with local departments in charge of water, electricity, transportation and meteorology, etc. Eight rounds of emergency water diversion to inhibit saltwater intrusion were conducted by the Commission, marking the most frequent responses and longest period of time than ever before. In total, 64 dispatch directives were issued and major reservoir group from the Xijiang and Beijiang rivers released 7.6 billion m^3 of water to the downstream. As a result, the average flow at the Wuhou Station was raised by almost 50%, and the freshwater/saltwater interface retreated by a maximum of about 41 km at the Modaomen waterway, thereby successfully checking the intrusion. Freshwater dispatched from the Pearl River also greatly contributed to the water storage downstream including Zhuhai and Zhongshan, 154 million m^3 of water was dispatched to Macao and Zhuhai. In particular, Macao SAR received 44 million m^3 of water from the reservoir group. It was the 19th year in a row that water from the Pearl River fed into the pipelines in the cities of the Greater Bay Area. This also generated multiple benefits, for example, maintaining uninterrupted waterway transport and safe operation of electric network in the Xijiang River, and bettering ecological and environmental conditions in the river channels.

案例 5　长江流域水库群蓄水调度

在 2022 年发生流域性特大干旱且连续两年流域来水总体偏少的情况下，长江委坚持旱涝同防同治，提前预留水源，以"蓄丰补枯"调度，为供水、生态、发电、航运等提供有力保障。2023 年汛前根据长江中下游水位大幅偏低及汛期旱重于涝的预测情况，科学调度三峡水库水位消落至 150 米左右，在三峡及金沙江梯级等水库留存了 30 多亿立方米水量，提前做好抗旱水源准备。汛期在确保防洪安全前提下，调度三峡水库水位基本稳定在 150 米以上运行，为流域城乡供水、农业灌溉、电力保供、航运畅通、生态良好提供水资源保障。长江委 8 月 30 日组织召开长江上游控制性水库群蓄水调度联合会商会，会同长江流域气象中心、交通运输部长江航务管理局、国家电网、南方电网、三峡集团、雅砻江公司等单位共同商议长江上游控制性水库群蓄水计划，及时编制《三峡水库蓄水计划》。及时审查批复了《金沙江下游梯级水库 2023 年联合蓄水计划》和瀑布沟、丹江口等水库的提前蓄水计划，指导水库群蓄水调度。加强与电网、发电企业的沟通协调，根据气象水文预测预报，抓住秋汛过程，统筹开展流域控制性水库群蓄水工作。9 月 10 日起三峡水库正式启动蓄水，起蓄水位较同期平均水位偏高 3.68 米，为完成年度蓄水任务争取了主动。10 月中下旬，三峡水库、丹江口水库、金中梯级水库、金下梯级水库、雅砻江两河口水库、大渡河瀑布沟水库均完成蓄水任务，其中三峡水库自 2010 年首次蓄满以来第 13 次蓄至正常蓄水位 175 米，丹江口水库第 2 次蓄至正常蓄水位 170 米。10 月 20 日，纳入联合调度的长江流域 53 座控制性水库死水位以上蓄水量 1069 亿立方米，自 2012 年长江流域水库群联合调度以来蓄水量首次超过 1000 亿立方米，其中，上游 29 座控制性水库死水位以上蓄水量 659 亿立方米，均创历史新高，为 2023 年冬 2024 年春枯水期补水提供了有效保障。

Case 5 Water storage and dispatch by reservoir group in the Yangtze River basin

Due to the extreme drought in the Yangtze basin in 2022 and less water inflow basinwide, the Changjiang Commission insisted on coordinating the prevention and control of both floods and droughts, reserved water resources in advance, and dispatched water following the principles of "restoring water during flooding and recharging water during low-water period", thus backing up water supply, river ecosystems, electricity generation and waterway transport. In 2023, based on the prediction before flooding that the water level in middle and lower reaches might be significantly low and drought might prevail over flood during flood seasons in the Yangtze River, the Commission scientifically drew down water level of the Three Gorges to about 150 m, and retained over 3 billion m³ of water in reservoirs such as the Three Gorges and the cascade reservoirs on the Jiansha River. During the flood season, while considering the safety of flood control, the Three Gorges was operated at over 150 m to guarantee the water resources for urban and rural water supply within the basin, agricultural irrigation, electricity supply, smooth waterway transport and sound ecosystems. On August 30, the Changjiang Commission convened a joint consultation meeting with the meteorological center of the Yangtze River basin, Changjiang River Administration of Navigational Affairs under the Ministry of Transport, State Grid, China Southern Power Grid, China Three Gorges Corporation, and Yalong Hydro, to discuss on water storage planning of controlling reservoir group in the upper Yangtze River and drafting of *The Impoundment Plan of the Three Gorges*. Relevant plans were reviewed and approved by the Commission to guide the water dispatch of reservoir group, including *the 2023 Plan on Joint Impoundment of Cascade Reservoirs in the Jinsha River Downstream*, and early impoundment plans of reservoirs such as Pubugou and Danjiangkou. Through close communication with companies leading electricity grid and power generation, the Commission fully took advantage of the autumn flooding based on meteorological and hydrological forecasting and prediction to initiate the impoundment of controlling reservoirs within the basin. On September 10, the Three Gorges officially started impoundment at a water level 3.68 m higher than normal, hence taking the initiative for achieving the annual goal of impoundment. In mid-to-late October, impoundment has been done in the Three Gorges, the Danjiangkou Reservoir, the cascade reservoirs in middle and lower Jinsha River, the Lianghekou Reservoir in the Yalong River and the Pubugou Reservoir in the Dadu River. It was the 13th time for the Three Gorges to reach a normal storage level at 175 m since the first full impoundment in 2010, and the second time for the Danjiangkou Reservoir to reach a normal storage level at 170 m. On October 20, 53 controlling reservoirs included in the joint dispatch action in the Yangtze basin impounded a 106.9 billion m³ of water above the dead storage level; it was also the first time since the joint dispatch of reservoir group began in 2012 that over 100 billion m³ was jointed stored in such manner. Particularly, 29 upstream controlling reservoirs stored 65.9 billion m³ above the dead storage level, marking the record high. It thus ensured the water replenishment during dry seasons in 2023 wintertime and 2024 springtime.

案例 6 甘肃省综合施策保供水

2023年，甘肃省河西地区遭遇60年一遇伏秋旱，民勤、永昌等县连续108天未出现5毫米以上降水，石羊河流域来水偏枯达8成，张掖、金昌、武威等地多数河道断流、水库干涸。旱情正值群众用水高峰期和秋粮生长关键期，14万人饮水受到影响，山丹县城46个小区5600多户群众分时供水，全省191千公顷作物受旱，其中16千公顷绝收。

一是"实"字为先，高位部署。甘肃省委、省政府12次召开会议，专题研究部署，主要领导一线指挥、现场推进，各部门协调联动、同频共振。水利厅及时启动干旱防御Ⅳ级应急响应，成立专班，连续125天对10市（州）30县（区）实行日调度，实时掌握旱情，科学研判形势，动态安排部署，并联合省委督查室和农业部门，派出7个工作组巡回指导、监督检查，确保抗旱措施落实落地。

二是"准"字为要，靶向发力。组织旱区水行政主管部门和水利工程管理单位，分级编制抗旱方案，算清"四本水账"（抗旱区域账、可用水源账、补水需求账、调水能力账），全面摸清供水、受水底数，分区域提出"五类措施"（取水引水、提水调水、水库补水、机井供水、节水限水），从地下找水、天上要水、山泉引水，因地制宜、以村施策，多措并举保供保灌保民生。

三是"抢"字当头，全力调水。水利部及黄委科学组织黄河、黑河抗旱调度，支持甘肃省2.95亿立方米抗旱应急水量。甘肃省水利厅组织景电、引大、引洮三大引调水工程深挖潜力、抢抓农时、全力调水，24小时满负荷运行，采取超常规措施加高加固"卡脖子段"，提升过流能力。三大工程用足抗旱应急水量，有力保障了受旱地区600多万群众饮水安全和800千公顷农田的灌溉需求。累计向水库、水池、塘坝补水1.65亿立方米，为冬春用水需求做足储备。

四是"保"字护航，开源节流。用好水利救灾资金，采取溪流截引、水源联通、延伸管网、集中供水、拉水送水、节水限水等多种方式，"组合出拳"，开源节流。抢抓中央企业支持新打441眼抗旱机井的机遇，派出5个专家组分赴5市12县（区），驻点服务，解决技术难题，有效缓解骨干工程覆盖区外170多万人的饮水安全问题。

Case 6 Integrated measures taken in Gansu Province to secure water supply

In 2023, the Hexi Area of Gansu Province went through a mid-summer and autumn drought of a 60-year return period. Several counties such as Minqin and Yongchang didn't receive any precipitation above 5 mm for 108 days in a row; the Shiyang river basin received 80% less inflow than normal; most of the river channels and reservoirs dried up in Zhangye, Jinchang and Wuwei. The drought struck during the peak season of residential water use and the critical growth period of autumn crops. As a result, 140,000 people's water use was affected, over 5,600 households from 46 residential compounds had tap water access in limited hours, 191,000 ha of cropland was stricken by drought and 16,000 ha of the affected cropland suffered crop failure.

First, prioritizing practical actions and making arrangements from the big picture. Gansu provincial government convened 12 meetings to analyze and discuss the conditions and solutions in response to droughts. The officials in charge commanded and supervised operations in the front line, various departments collaborated closely with a clear focus. The provincial water department promptly launched a Level IV emergency response against drought and established a dedicated working group. The working group conducted daily assessment in 10 cities/autonomous prefectures and 30 counties/districts for over 125 days in a row, monitoring the drought development and making timely decisions. Along with various other departments like the agricultural department, the provincial government sent out seven working groups for guidance, supervision and inspection to ensure effective implementation of drought relief measures.

Second, a clear focus with targeted measures. The local government mobilized water-related administrative departments and engineering management units in drought-affected areas to draft drought relief plans. Calculations were made on the "Four water accounts" including regional water resources for drought relief, available water sources, water demands for combating shortage and water transfer and diversion capacity, so as to comprehensively assess water supply and demand. "Five aspects of measures" were proposed for different regions, regarding water extraction, water diversion, reservoir recharge, water supply via pumping wells and water conservation and rationing. Various water sources such as groundwater, rainwater, mountain springs and strategies tailored to the needs of every village were used to ensure water supply, irrigation use, and livelihood.

Third, focusing on emergency water diversion. MWR and the Yellow River Commission scientifically organized drought relief operations in the Yellow River and the Heihe River, providing 295 million m³ of emergency water to Gansu Province. Three major water diversion projects including the Jingtaichuan electrical water lifting for irrigation project, the Dadu-Minjiang water diversion project and the Taohe River water diversion project were operated at full capacity 24 hours for water diversion and for meet the seasonal irrigation demands; extraordinary measures were taken to heighten and reinforce the narrowest parts of the discharge channel to increase flow passage. Emergency water dispatched from the three projects effectively guaranteed drinking water needs of over 6 million residents and irrigation water needs of 800,000 ha of cropland in the drought-affected areas. A total of 165 million m³ of water was replenished to reservoirs and small ponds reservoirs in various sizes, thereby preparing for water demands in winter and next spring.

Last but not least, diversifying water sources to secure water supply. Water-related disaster relief fund was wisely used to develop new water sources and conserve the existent ones: water diversion from streams, connecting different water sources, extending pipelines, centralizing water supply, water transport and delivery and water saving, etc. During the period when 441 new pumping wells were drilled under the support of state-owned companies, Gansu dispatched five expert teams to 12 counties/districts in five cities to provide on-site technological support, greatly contributing to the drinking water security for over 1.7 million people in the areas that could not benefit from the major water projects.

4 基础工作
FOUNDATIONAL WORK

4.1 机构职能

2023年12月12日，国家防汛抗旱总指挥部批复同意将辽河流域防汛抗旱协调领导小组调整为辽河防汛抗旱总指挥部，日常办事机构设在松辽委。至此，长江、黄河、淮河、海河、珠江、松花江、辽河七大江河及太湖流域防汛抗旱指挥机构全部建立。

4.1 Institutional Functions

On December 12, 2023, the National Flood Control and Drought Relief Headquarters approved the reorganization of the Liaohe River Basin Flood and Drought Coordination and Leading Group into the Liaohe River Flood and Drought Relief Headquarters, with the daily operational office set up in the Songliao Commission. This marks the completion of establishing FDHs for the seven major river basins, including the Yangtze, Yellow, Huaihe, Haihe, Pearl, Songhua, and Liaohe rivers, as well as the Taihu Lake Basin.

4.2 规章制度

水利部印发《长江流域控制性水工程联合调度管理办法（试行）》《江河湖库旱警水位（流量）管理规定（试行）》《流域防洪工程联合调度方案编制导则》，组织修订《山洪灾害防御预案编制技术导则》。黄委修订《黄河防汛抗旱应急预案（试行）》。淮委修编印发《淮河防汛抗旱总指挥部防汛抗旱应急预案》《淮河流域水情旱情预警发布管理办法》。海委编制《漳卫河系防汛协调联动工作机制》。珠江委编制印发《珠江流域（片）水旱灾害防御应急响应工作规程》。

4.2 Rules and Regulations

The MWR issued the *Regulations for Joint Dispatch and Management of Key Control Projects in the Yangtze River Basin (Trial)*, *Regulations on Drought Warning Water Levels for Rivers, Lakes, and Reservoirs (Trial)*, and *Guidelines for the Compilation of Joint Dispatch Plans for Flood Control Projects in River Basins*. It also organized the revision of *The Technical Guidelines for the Compilation of Flash Flood Disaster Contingency Plans*. The Yellow River Commission revised *The Emergency Response Plan for Flood and Drought Disaster Prevention (Trial)*. The Huaihe Commission revised and issued *The Emergency Response Plan for Flood and Drought Prevention by the Huaihe River Flood and Drought Relief Headquarters* and *Management Measures for the Issuance of Flood and Drought Warnings in the Huaihe River Basin*. The Haihe Commission compiled *The Coordination Mechanism of Flood Control for the Zhang-Wei River System*. The Pearl River Commission drafted and issued *The Emergency Response Procedures for Flood and Drought Disaster Prevention in the Pearl River Basin*.

4.3 方案预案

水利部编制《水利部贯彻落实习近平总书记关于未雨绸缪做好农业防灾减灾重要指示实施方案》，批复《2023年长江流域水工程联合调度运用计划》《2023年雄安新区起步区安全度汛方案》《2023年珠江流域汛期水库群联合调度运用计划》《太湖流域洪水与水量调度方案》《太湖流域水资源调度方案》，组织七大流域编制2023年度抗旱应急水量调度预案。长江委编制《贯彻落实习近平总书记重要指示做好2023年主汛期水旱灾害防御重点工作方案》。黄委编制修订《2023年黄河中下游洪水调度方案》《2023年黄河上游重要水库群联合防洪调度方案》《2023年黄河调水调沙预案》《2023年黄河中下游洪水应急调度方案》《沁河应急水量调度方案》，批复11座委直接调度管理的大型水库年度调度运用计划。淮委编制修订《淮河大型水库群联合调度方案》，审核修订流域24处蓄滞洪区运用预案及主要河道、大型水库调度运用计划。海委修订滦河、北三河、永定河、大清河、子牙河、漳卫河超标洪水防御预案，编制东张务湿地蓄滞洪区调度运用方案。松辽委编制《2023年松花江流域水工程防洪联合调度方案》《2023年辽河流域水工程防洪联合调度方案》，批复尼尔基、察尔森、丰满、白山4座骨干水库2023年汛期调度运用计划。太湖局编制修订《太湖流域骨干工程防洪调度预警工作方案》《太湖流域2023年后汛期防洪蓄水方案》。

4.3 Contingency Planning

The MWR drafted *The Implementation Plan on Executing the Key Central Instructions on Proactive Disaster Prevention and Reduction in Agriculture*, approved *The 2023 Joint Scheduling and Operation Plan for Water Projects in the Yangtze River Basin*, *The 2023 Flood Safety Plan for the Start-up Zone of Xiong'an New Area*, *The 2023 Joint Scheduling and Operation Plan for Reservoir Groups During the Flood Season in the Pearl River Basin*, *The Dispatch Plan for Flood and Water in the Taihu Basin*, and *The Water Resources Regulation and Dispatch Plan for Taihu Basin*. It organized the compilation of the 2023 Emergency Water Regulation and Dispatch Plans for Drought Relief for the seven major basins.

The Changjiang Commission drafted *The Plan for Key Tasks in Flood and Drought Disaster Prevention During the 2023 Flood Season*.

The Yellow River Commission amended and formulated *The 2023 Floodwater Regulation Plan for the Middle and Lower Reaches of the Yellow River*, *The 2023 Joint Flood Control Plan for Major Reservoirs in the Upper Yellow River*, *The 2023 Flow and Sediment Regulation Plan for Yellow River*, *The 2023 Contingency Plan for Flood Dispatch and Control in the Middle and Lower Yellow River*, and *The Emergency Water Dispatch Plan of the Qinhe River*, and approved the annual operation plans for 11 large reservoirs directly managed by the Commission.

The Huaihe Commission drafted and revised *The Joint Dispatch and Operation Plan for Major Reservoir Groups of the Huaihe River*. It also reviewed and amended the operation plans for 24 flood detention areas, and major river channels and large reservoirs.

The Haihe Commission revised flood prevention plans for extreme floods in the Luanhe River, Beisan River, Yongding River, Daqing River, Ziya River, and Zhang-Wei River, and compiled operation plan for the flood detention area of Dongzhangwu Wetland.

The Songliao Commission developed *The 2023 Joint Scheduling Plan of Water Projects for Flood Prevention and Control in the Songhua River Basin* and *The 2023 Joint Scheduling Plan of Water Projects for Flood Prevention and Control in the Liaohe River Basin*, and approved the flood season operation plans for the backbone reservoirs of Nierji, Chaersen, Fengman, and Baishan.

The Taihu Authority drafted and revised *The Dispatch and Early Warning Plan for Flood Prevention and Control by Major Water Projects of the Taihu Basin* and *The 2023 Taihu Basin Flood Control and Impoundment Plan during the Late Flood Season*.

4.4 山洪灾害防治

水利部会同财政部下达中央水利发展资金 24 亿元，支持地方开展山洪灾害补充调查评价、监测预警能力巩固提升、简易监测预警实施设备配备、小流域山洪灾害"四预"能力建设等山洪灾害防治非工程措施建设及运行维护，实施 195 条重点山洪沟防洪治理，加快完善非工程措施与工程措施相结合的山洪灾害防御体系。积极争取 2023 年增发国债资金 175.8 亿元，实施 1891 条重点山洪沟防洪治理，持续提升沿河村镇和重要基础设施防冲能力。先后在浙江、安徽 2 次组织召开项目建设管理工作会议、视频调度会议，部署规范项目建设管理工作。

4.4 Flash Flood Disaster Management

The MWR, in conjunction with the Ministry of Finance, allocated 2.4 billion RMB from the central water development funds to support local efforts in supplementary investigations and assessments of flash flood disasters, consolidation and enhancement of monitoring and early warning capacity, provision of simple monitoring and warning equipment, and the development of the four preemptive pillars for small basin flash flood disasters. These funds also supported the construction and maintenance of non-engineering facilities, and the implementation of flood control measures in 195 key gullies prone to flash floods, accelerating the refining of a prevention and control system that integrates non-engineering and engineering measures. An additional 17.58 billion RMB from treasury bond was secured in 2023 to implement flood control measures in 1,891 gullies prone to flash floods, and to enhance the resilience of riverside villages and key infrastructure to floods. Meetings and video conferences were organized in Zhejiang and Anhui Provinces to standardize the construction and management of such projects.

4.5 蓄滞洪区建设管理

水利部积极推动国家蓄滞洪区建设项目前期工作，江西省鄱阳湖康山、珠湖、黄湖、方洲斜塘，安徽省淮河流域一般行蓄洪区等3个项目开工建设。加快推进国家蓄滞洪区在建项目实施，2023年共下达中央预算内投资计划99.25亿元（其中中央投资59.28亿元），支持大陆泽、宁晋泊、杜家台、华阳河等12个国家蓄滞洪区建设项目。其中，安徽省淮河行蓄洪区和淮河干流滩区居民迁建（2023年度）、淮河干流正阳关至峡山口段行洪区调整和建设、淮河干流王家坝至临淮岗段行洪区调整及河道整治、广东省湛江蓄滞洪区建设与管理等4个项目基本完工。强化蓄滞洪区运行管理，对98处国家蓄滞洪区逐一建档立卡，逐一明确建设管理目标任务，逐一开展安全运用分析评价，完善数字一张图，形成"一区一册"。商财政部首次将国家蓄滞洪区工程维修养护纳入中央财政水利发展资金支持范围，支持国家蓄滞洪区约6000千米堤防和110多座进退洪闸的维修养护。

4.5 Construction and Management of Flood Detention Areas

The MWR actively promoted the preliminary work for the construction of national flood detention areas. Construction commenced on these projects: Kangshan, Zhuhu, Huanghu, Fangzhouxietang surrounding the Poyang Lake (Jiangxi Province), and the regular flood detention area in the Huaihe River Basin (Anhui Province). Efforts were intensified to advance the completion of ongoing national flood detention area projects, with a total of 9.925 billion RMB allocated from the central government budget in 2023 (including 5.928 billion RMB of central government funding), benefiting the 12 national flood detention areas of Daluze, Ningjinpo, Dujiatai, and Huayang River. Four projects were essentially completed: the relocation of residents from the flood detention areas and the floodplain of the Huaihe River (Anhui Province, 2023), the adjustment and construction of the flood detention area for the Zhengyangguan-Xiashankou section of the Huaihe River (Anhui Province), the flood detention area adjustment and river training and improvement for from the Wangjiaba-Linhuaigang section of the Huaihe River (Anhui Province), and the construction and management of the Pajiang River flood detention area (Guangdong Province). The operation and management of flood detention areas were strengthened, with records established for each of the 98 basins, construction and management goals set for each, and safety operation analyses and evaluations conducted for each. In addition, better digital mapping was conducted for each basin to form a "one basin, one dossier" system. For the first time, the Ministry of Finance included the infrastructure repair and maintenance in national flood detention areas within the scope of central financial support for water conservancy development, financing the maintenance of approximately 6,000 km of levees and over 110 floodgates within the basins.

4.6 信息发布

水利部坚持正面宣传、积极主动发声，李国英部长在国务院新闻办公室举行"权威部门话开局"系列主题新闻发布会上介绍水旱灾害防御有关情况；针对海河"23·7"流域性特大洪水召开新闻发布会，举办《扛牢天职 确保安全——海河"23·7"流域性特大洪水防御工作》图片展；全年编发《汛旱情通报》114期，及时发布水旱灾害防御相关信息。强化与《人民日报》、新华社等中央主要媒体重要汛情和防御情况通报机制，防汛关键期与央视建立每日通报机制，《新闻联播》连续13天报道水利部防汛工作情况，《焦点访谈》《东方时空》《新闻1+1》等栏目对海河流域洪水防御进行专题解读；水利部水旱灾害防御司、信息中心、中国水利水电科学研究院、相关流域管理机构专家和水利部工作组接受媒体采访60余人次，主动回应社会关切，营造良好舆论氛围。加强水旱灾害防御科普宣传，配合央视制作科普产品《涿州的水到底从哪来》，通过水利部官微发布《为何暴雨"突如其来"？遇极端降水如何避险？》《今日入汛，这些"汛"息很重要！》等科普图文，普及防洪避险知识。

4.6 Information Dissemination

The MWR adhered to proactive and positive public communication strategies. Minister Li Guoying introduced relevant measures for flood and drought disaster prevention at a press conference held by the State Council Information Office as part of the "Ministers on the Kickoff of 2023" series. In response to "23·7" extreme floods of the Haihe River basin, a press conference was convened, and a photo exhibition titled *Upholding Our Duty, Ensuring Safety—Flood Control during "23·7" Extreme Flood* was organized. Throughout the year, 114 volumes of *the Flood and Drought Newsletter* were compiled and issued, disseminating information related to prevention and control of flood and drought disasters. The Ministry strengthened the reporting mechanism of important news on flood control with major central media outlets such as *People's Daily* and Xinhua News Agency. During critical periods of flood control, a daily reporting mechanism was established with China Central Television (CCTV). *Xinwen Lianbo*, the prime time news program, reported on the Ministry's flood control activities for 13 consecutive days. *Programs such as Focus*, *Oriental Horizon*, and *News 1+1* offered themed discussions on flood control of the Haihe River Basin. Experts from the Department of Flood and Drought Disaster Prevention, the Information Center, the China Institute of Water Resources and Hydropower Research, relevant river basin commissions (authorities), and working groups of the MWR engaged in over 60 media interviews, actively responding to public concerns and fostering a positive public opinion atmosphere. The Ministry also intensified public education on flood and drought disaster prevention, cooperating with CCTV to produce the science education program *Unveiling the Source: What Causes Zhuozhou's Floods?* and disseminating infographics and texts such as *Sudden Downpours Explained: What Causes Unexpected Heavy Rains?* and *Flood Season Alert: Essential Facts You Need to Know Today!* through its WeChat subscription account, spreading knowledge about flood prevention and risk avoidance.

4.7 复盘分析

水利部坚持问题导向、结果导向，全面系统调查分析灾害发生过程，开展复盘分析工作。以雨水情测报、关键水利工程调度、蓄滞洪区运用等为重点开展海河流域洪水防御工作复盘分析，全面检视洪水防御各环节工作，总结防御工作经验，剖析存在的问题，提出改进措施。复核松花江支流拉林河设计洪水，全面检视磨盘山、龙凤山水库防洪调度过程，查找水库调度链条上存在的问题和不足，针对性地提出改进建议。全面复盘检视四川汶川、金阳，重庆万州，浙江富阳，陕西长安，甘肃夏河等地山洪（或伴生泥石流）灾害事件"四预"工作，科学分析致灾原因，认真查找薄弱环节，研究提出完善山洪灾害防御体系的对策措施。开展大汶河流域暴雨洪水防御调查评估工作，着重雨情、水情、工情调查及洪水预报调度复盘分析，检视预报调度工作及数字孪生流域建设效果。

4.7 Review and Analysis

The MWR, adopting a problem-oriented and results-driven approach, conducted comprehensive and systematic investigations and analyses of disaster processes in retrospect. A detailed assessment of flood control operations in the Haihe River Basin was undertaken, focusing on rainfall and water regimes forecast, coordinated operation of major water conservancy projects, and the utilization of flood detention areas. This analysis scrutinized all facets of flood control efforts, summarized best and worst practices, identified current issues, and suggested improvements. A re-evaluation of flooding in the Lalin River, a tributary of the Songhua River, was performed, alongside a detailed review of flood dispatch processes at the Mopanshan and Longfengshan Reservoirs. This review pinpointed flaws and deficiencies in the reservoir dispatching chain and recommended specific enhancements. Comprehensive reviews were also carried out regarding the four preemptive measures against flash floods and associated debris flow disasters in several locations, including Wenchuan and Jinyang in Sichuan Province, Wanzhou in Chongqing City, Fuyang in Zhejiang Province, Chang'an in Shaanxi Province, and Xiahe in Gansu Province. These analyses scientifically explored the causes of disasters, identified vulnerabilities, and proposed measures to bolster the flash flood prevention system. Another focus was the intense rainfall and flood control efforts in the Dawen River Basin. Emphasis was placed on investigating precipitation and water conditions, infrastructure status, and on reviewing and evaluating flood forecasting and dispatch processes as well as the effectiveness of digital twinning of the basin.

附录 APPENDIX

1950—2023 年全国水旱灾情统计

STATISTICS OF FLOOD AND DROUGHT DISASTERS IN CHINA 1950-2023

附表1 1950—2023年全国洪涝灾情统计
Appendix 1　Flood disasters and losses 1950-2023

年份 Year	农作物受灾面积/ 千公顷 Affected cropland area/1,000 ha	农作物成灾面积/ 千公顷 Failed cropland area/1,000 ha	因灾死亡人口/ 人 Deaths/person	因灾失踪人口/ 人 Missing persons/person	倒塌房屋/ 万间 Collapsed dwellings/10,000 rooms	直接经济损失/ 亿元 Direct economic loss/100 million RMB
1950	6559.00	4710.00	1982	—	130.50	—
1951	4173.00	1476.00	7819	—	31.80	—
1952	2794.00	1547.00	4162	—	14.50	—
1953	7187.00	3285.00	3308	—	322.00	—
1954	16131.00	11305.00	42447	—	900.90	—
1955	5247.00	3067.00	2718	—	49.20	—
1956	14377.00	10905.00	10676	—	465.90	—
1957	8083.00	6032.00	4415	—	371.20	—
1958	4279.00	1441.00	3642	—	77.10	—
1959	4813.00	1817.00	4540	—	42.10	—
1960	10155.00	4975.00	6033	—	74.70	—
1961	8910.00	5356.00	5074	—	146.30	—
1962	9810.00	6318.00	4350	—	247.70	—
1963	14071.00	10479.00	10441	—	1435.30	—
1964	14933.00	10038.00	4288	—	246.50	—
1965	5587.00	2813.00	1906	—	95.60	—
1966	2508.00	950.00	1901	—	26.80	—
1967	2599.00	1407.00	1095	—	10.80	—
1968	2670.00	1659.00	1159	—	63.00	—
1969	5443.00	3265.00	4667	—	164.60	—
1970	3129.00	1234.00	2444	—	25.20	—
1971	3989.00	1481.00	2323	—	30.20	—
1972	4083.00	1259.00	1910	—	22.80	—
1973	6235.00	2577.00	3413	—	72.30	—

续表 Continued

年份 Year	农作物受灾面积/千公顷 Affected cropland area/1,000 ha	农作物成灾面积/千公顷 Failed cropland area/1,000 ha	因灾死亡人口/人 Deaths/person	因灾失踪人口/人 Missing persons/person	倒塌房屋/万间 Collapsed dwellings/10,000 rooms	直接经济损失/亿元 Direct economic loss/100 million RMB
1974	6431.00	2737.00	1849	—	120.00	—
1975	6817.00	3467.00	29653	—	754.30	—
1976	4197.00	1329.00	1817	—	81.90	—
1977	9095.00	4989.00	3163	—	50.60	—
1978	2820.00	924.00	1796	—	28.00	—
1979	6775.00	2870.00	3446	—	48.80	—
1980	9146.00	5025.00	3705	—	138.30	—
1981	8625.00	3973.00	5832	—	155.10	—
1982	8361.00	4463.00	5323	—	341.50	—
1983	12162.00	5747.00	7238	—	218.90	—
1984	10632.00	5361.00	3941	—	112.10	—
1985	14197.00	8949.00	3578	—	142.00	—
1986	9155.00	5601.00	2761	—	150.90	—
1987	8686.00	4104.00	3749	—	92.10	—
1988	11949.00	6128.00	4094	—	91.00	—
1989	11328.00	5917.00	3270	—	100.10	—
1990	11804.00	5605.00	3589	—	96.60	239.00
1991	24596.00	14614.00	5113	—	497.90	779.08
1992	9423.30	4464.00	3012	—	98.95	412.77
1993	16387.30	8610.40	3499	—	148.91	641.74
1994	18858.90	11489.50	5340	—	349.37	1796.60
1995	14366.70	8000.80	3852	—	245.58	1653.30
1996	20388.10	11823.30	5840	—	547.70	2208.36
1997	13134.80	6514.60	2799	—	101.06	930.11
1998	22291.80	13785.00	4150	—	685.03	2550.90
1999	9605.20	5389.12	1896	—	160.50	930.23

附录　1950—2023 年全国水旱灾情统计

续表 Continued

年份 Year	农作物受灾面积 / 千公顷 Affected cropland area/1,000 ha	农作物成灾面积 / 千公顷 Failed cropland area/1,000 ha	因灾死亡人口 / 人 Deaths/person	因灾失踪人口 / 人 Missing persons/person	倒塌房屋 / 万间 Collapsed dwellings/10,000 rooms	直接经济损失 / 亿元 Direct economic loss/100 million RMB
2000	9045.01	5396.03	1942	–	112.61	711.63
2001	7137.78	4253.39	1605	–	63.49	623.03
2002	12384.21	7439.01	1819	–	146.23	838.00
2003	20365.70	12999.80	1551	–	245.42	1300.51
2004	7781.90	4017.10	1282	–	93.31	713.51
2005	14967.48	8216.68	1660	–	153.29	1662.20
2006	10521.86	5592.42	2276	–	105.82	1332.62
2007	12548.92	5969.02	1230	–	102.97	1123.30
2008	8867.82	4537.58	633	232	44.70	955.44
2009	8748.16	3795.79	538	110	55.59	845.96
2010	17866.69	8727.89	3222	1003	227.10	3745.43
2011	7191.50	3393.02	519	121	69.30	1301.27
2012	11218.09	5871.41	673	159	58.60	2675.32
2013	11777.53	6540.81	775	374	53.36	3155.74
2014	5919.43	2829.99	486	91	25.99	1573.55
2015	6132.08	3053.84	319	81	15.23	1660.75
2016	9443.26	5063.49	686	207	42.77	3643.26
2017	5196.47	2781.19	316	39	13.78	2142.53
2018	6426.98	3131.16	187	32	8.51	1615.47
2019	6680.40	3928.97	573	85	10.30	1922.70
2020	7190.00	4118.21	230	49	9.00	2669.80
2021	4760.43	2643.05	512	78	15.20	2458.92
2022	3413.73	1834.57	143	28	3.13	1288.99
2023	4633.29	2320.90	309		13.00	2445.75

注 2019—2023 年数据来源于应急管理部；"–" 表示没有统计数据；因灾失踪人口从 2008 年开始作为指标统计。
Note Data during 2019−2023 are from the Ministry of Emergency Management; "–" means statistics don't exist; missing persons attributed to disasters was determined as a statistical indicator since 2008.

附表2 1950—2023年全国干旱灾情统计
Appendix 2　Drought disasters and losses 1950-2023

年份 Year	农作物因旱受灾面积／千公顷 Affected cropland area/1,000 ha	农作物因旱成灾面积／千公顷 Damaged cropland area/1,000 ha	农作物因旱绝收面积／千公顷 Area of crop failure/1,000 ha	因旱粮食损失／亿千克 Crop losses/100 million kg	因旱饮水困难人口／万人 People with drinking water difficulties/10,000 persons	因旱饮水困难大牲畜／万头 Number of bigger-sized livestock having difficulties accessing drinking water/10,000 heads	直接经济损失／亿元 Direct economic loss/100 million RMB
1950	2398.00	589.00	—	19.00	—	—	—
1951	7829.00	2299.00	—	36.88	—	—	—
1952	4236.00	2565.00	—	20.21	—	—	—
1953	8616.00	1341.00	—	54.47	—	—	—
1954	2988.00	560.00	—	23.44	—	—	—
1955	13433.00	4024.00	—	30.75	—	—	—
1956	3127.00	2051.00	—	28.60	—	—	—
1957	17205.00	7400.00	—	62.22	—	—	—
1958	22361.00	5031.00	—	51.28	—	—	—
1959	33807.00	11173.00	—	108.05	—	—	—
1960	38125.00	16177.00	—	112.79	—	—	—
1961	37847.00	18654.00	—	132.29	—	—	—
1962	20808.00	8691.00	—	89.43	—	—	—
1963	16865.00	9021.00	—	96.67	—	—	—
1964	4219.00	1423.00	—	43.78	—	—	—
1965	13631.00	8107.00	—	64.65	—	—	—
1966	20015.00	8106.00	—	112.15	—	—	—
1967	6764.00	3065.00	—	31.83	—	—	—
1968	13294.00	7929.00	—	93.92	—	—	—
1969	7624.00	3442.00	—	47.25	—	—	—
1970	5723.00	1931.00	—	41.50	—	—	—
1971	25049.00	5319.00	—	58.12	—	—	—
1972	30699.00	13605.00	—	136.73	—	—	—
1973	27202.00	3928.00	—	60.84	—	—	—
1974	25553.00	2296.00	—	43.23	—	—	—
1975	24832.00	5318.00	—	42.33	—	—	—

年份 Year	农作物因旱受灾面积/千公顷 Affected cropland area/1,000 ha	农作物因旱成灾面积/千公顷 Damaged cropland area/1,000 ha	农作物因旱绝收面积/千公顷 Area of crop failure/1,000 ha	因旱粮食损失/亿千克 Crop losses/100 million kg	因旱饮水困难人口/万人 People with drinking water difficulties/10,000 persons	因旱饮水困难大牲畜/万头 Number of bigger-sized livestock having difficulties accessing drinking water/10,000 heads	直接经济损失/亿元 Direct economic loss/100 million RMB
1976	27492.00	7849.00	–	85.75	–	–	–
1977	29852.00	7005.00	–	117.34	–	–	–
1978	40169.00	17969.00	–	200.46	–	–	–
1979	24646.00	9316.00	–	138.59	–	–	–
1980	26111.00	12485.00	–	145.39	–	–	–
1981	25693.00	12134.00	–	185.45	–	–	–
1982	20697.00	9972.00	–	198.45	–	–	–
1983	16089.00	7586.00	–	102.71	–	–	–
1984	15819.00	7015.00	–	106.61	–	–	–
1985	22989.00	10063.00	–	124.04	–	–	–
1986	31042.00	14765.00	–	254.34	–	–	–
1987	24920.00	13033.00	–	209.55	–	–	–
1988	32904.00	15303.00	–	311.69	–	–	–
1989	29358.00	15262.00	2423.33	283.62	–	–	–
1990	18174.67	7805.33	1503.33	128.17	–	–	–
1991	24914.00	10558.67	2108.67	118.00	4359.00	6252.00	–
1992	32980.00	17048.67	2549.33	209.72	7294.00	3515.00	–
1993	21098.00	8658.67	1672.67	111.80	3501.00	1981.00	–
1994	30282.00	17048.67	2526.00	233.60	5026.00	6012.00	–
1995	23455.33	10374.00	2121.33	230.00	1800.00	1360.00	–
1996	20150.67	6247.33	686.67	98.00	1227.00	1675.00	–
1997	33514.00	20010.00	3958.00	476.00	1680.00	850.00	–
1998	14237.33	5068.00	949.33	127.00	1050.00	850.00	–
1999	30153.33	16614.00	3925.33	333.00	1920.00	1450.00	–
2000	40540.67	26783.33	8006.00	599.60	2770.00	1700.00	–
2001	38480.00	23702.00	6420.00	548.00	3300.00	2200.00	–

续表 Continued

年份 Year	农作物因旱受灾面积/千公顷 Affected cropland area/1,000 ha	农作物因旱成灾面积/千公顷 Damaged cropland area/1,000 ha	农作物因旱绝收面积/千公顷 Area of crop failure/1,000 ha	因旱粮食损失/亿千克 Crop losses/100 million kg	因旱饮水困难人口/万人 People with drinking water difficulties/10,000 persons	因旱饮水困难大牲畜/万头 Number of bigger-sized livestock having difficulties accessing drinking water/10,000 heads	直接经济损失/亿元 Direct economic loss/100 million RMB
2002	22207.33	13247.33	2568.00	313.00	1918.00	1324.00	-
2003	24852.00	14470.00	2980.00	308.00	2441.00	1384.00	-
2004	17255.33	7950.67	1677.33	231.00	2340.00	1320.00	-
2005	16028.00	8479.33	1888.67	193.00	2313.00	1976.00	-
2006	20738.00	13411.33	2295.33	416.50	3578.23	2936.25	986.00
2007	29386.00	16170.00	3190.67	373.60	2756.00	2060.00	1093.70
2008	12136.80	6797.52	811.80	160.55	1145.70	699.00	545.70
2009	29258.80	13197.10	3268.80	348.49	1750.60	1099.40	1206.59
2010	13258.61	8986.47	2672.26	168.48	3334.52	2440.83	1509.18
2011	16304.20	6598.60	1505.40	232.07	2895.45	1616.92	1028.00
2012	9333.33	3508.53	373.80	116.12	1637.08	847.63	533.00
2013	11219.93	6971.17	1504.73	206.36	2240.54	1179.35	1274.51
2014	12271.70	5677.10	1484.70	200.65	1783.42	883.29	909.76
2015	10067.05	5577.04	1005.39	144.41	836.43	806.77	579.22
2016	9872.76	6130.85	1018.20	190.64	469.25	649.73	484.15
2017	9946.43	4490.02	752.71	134.44	477.78	514.29	437.88
2018	7397.21	3667.23	610.21	156.97	306.69	462.30	483.62
2019	7838.00	4760.17	1113.60	92.29	692.29	368.10	457.40
2020	5081.00	2759.08	704.50	123.04	668.98	448.63	249.20
2021	3426.16	1949.00	464.12	49.28	546.35	250.63	200.87
2022	6090.21	2858.39	611.78	57.44	542.38	331.92	512.85
2023	3803.70	1488.90	218.37	33.88	274.74	278.29	205.51

注 2019—2023 年数据第 2、3、4、8 列来源于应急管理部；第 2 列 "农作物因旱受灾面积" 2019 年之前为 "作物因旱受灾面积"；第 3 列 "农作物因旱成灾面积" 2019 年之前为 "作物因旱成灾面积"；第 4 列 "农作物因旱绝收面积" 2019 年之前为 "作物因旱绝收面积"；"-" 表示没有统计数据。

Note Data during 2019-2023 in columns 2, 3, 4 & 8 are from the Ministry of Emergency Management; "-" means statistics don't exist.